Prisoners of Space?

Westview Replica Editions

This book is a Westview Replica Edition. The concept of Replica Editions is a response to the crisis in academic and informational publishing. Library budgets for books have been severely curtailed; economic pressures on the university presses and the few private publishing companies primarily interested in scholarly manuscripts have severely limited the capacity of the industry to properly serve the academic and research communities. Many manuscripts dealing with important subjects, often representing the highest level of scholarship, are today not economically viable publishing projects. Or, if they are accepted for publication, they are often subject to lead times ranging from one to three years. Scholars are understandably frustrated when they realize that their first-class research cannot be published within a reasonable time frame, if at all.

Westview Replica Editions are our practical solution to the problem. The concept is simple. We accept a manuscript in camera-ready form and move it immediately into the production process. The responsibility for textual and copy editing lies with the author or sponsoring organization. If necessary we will advise the author on proper preparation of footnotes and bibliography. We prefer that the manuscript be typed according to our specifications, though it may be acceptable as typed for a dissertation or prepared in some other clearly organized and readable way. The end result is a book produced by lithography and bound in hard covers. Initial edition sizes range from 600 to 800 copies, and a number of recent Replicas are already in second printings. We include among Westview Replica Editions only works of outstanding scholarly quality or of great informational value, and we will continue to exercise our usual editorial standards and quality control.

Prisoners of Space?: Exploring the Geographical Experience of Older People
Graham D. Rowles

This book suggests that the older person's environmental experience is far more complex than the prevalent societal image of progressive spatial constriction with advancing years. The study is an in-depth exploration of the geographical experience, defined as "involvement within the spaces and places of their lives," of five elderly persons who have lived for many years in a working-class inner city neighborhood. Over a two-year period, the author developed close relationships with the participants, and through this intimacy discovered many of the nuances of their experiences.

Each person's geographical experience is described in a vignette that reveals the essential features of his or her life-style. Stan's stoic resignation, Marie's aggressive defiance, Raymond's jovial acceptance, Evelyn's placid equanimity, and Edward's calm accommodation illustrate five highly individual responses to growing old. Each response is reflected in a unique relationship with the shared environmental context. However, closer scrutiny reveals common underlying themes, among them a vicarious participation in places of the past and in distant locations such as the homes of the subjects' children. The author suggests that constriction in the realm of action is attended by expansion of the role of geographical fantasy together with consistent changes in orientation within space and in the individual's feelings about the places of his or her life. A number of policy implications of this perspective are explored.

Graham D. Rowles, assistant professor at West Virginia University, received his early schooling in England and his Ph.D. in geography from Clark University. Much of his work has focused on the way elderly people experience the urban environment, and he is now exploring the environmental experience of the Appalachian elderly.

. . . but inside this old carcass a young girl still dwells
And now and again my battered heart swells.
I remember the joys, I remember the pain
And I'm loving and living life over again.
I think of the years, all too few—gone too fast,
And accept the stark fact that nothing can last.
So open your eyes, nurses, open and see
Not a crabbit old woman, look closer—
See me.

Written by an old woman
who died, quietly and alone,
in the geriatric ward of an
Irish hospital.

Prisoners of Space?
Exploring the Geographical Experience of Older People

Graham D. Rowles

To Nancy, my friend and colleague:

Expressing appreciation for shared visions of scholarship without disciplinary limits, experience beyond intellect and friendship transcending mere consensus.

—with love

Graham

Westview Press • Boulder, Colorado

A Westview Replica Edition

Copyright © 1978 by Westview Press, Inc.

Published in 1978 in the United States of America by
  Westview Press, Inc.
  5500 Central Avenue
  Boulder, Colorado 80301
  Frederick A. Praeger, Publisher

Paperback edition published in September 1980 by Westview Press, Inc.

Library of Congress Cataloging in Publication Data
Rowles, Graham D.
  The prisoners of space?
  (A Westview replica edition)
  Bibliography: p.
  1. Aged—United States—Case studies. I. Title. II. Title: Geographical experience of older people.
HQ1064.U5R68   301.43'5'0973   77-18655
ISBN 0-89158-069-7
ISBN 0-86531-072-6 pbk

Printed and bound in the United States of America

For Stan, Marie, Raymond, Evelyn, and Edward.

CONTENTS

CHAPTER VIII

FIGURES

PLATES

TABLE

ACKNOWLEDGMENTS

The completion of this work is testimony to the tolerance and sensitivity of a large number of people who unflinchingly committed themselves to what, from the outset, was clearly a high risk venture. Unfortunately, Stan and Edward are no longer alive to read these words, but to Marie, Raymond, and Evelyn, I would like to express thanks for allowing me to become a part of their lives. Without their tolerance of an imposing, probing and yet somewhat confused "youth" this study would not have been possible.

Thanks are also due to friends in several academic settings for their forbearance in listening to my lengthy verbal ramblings on half formed ideas, or responding in writing to early drafts. Specifically, I am indebted to Charles Blinderman of the Clark University English Department; Joey Edwardh of the Department of Geography, Syracuse University; and David Seamon, Mick Godkin, Curt Rose and Marc Eichen, former colleagues in the Graduate School of Geography at Clark. Marc in particular invested many hours in the venture.

This book was originally written as a doctoral dissertation. Clearly, the contribution of research advisors is a crucial element in the successful completion of any dissertation. I was fortunate in having a committee providing a challenging blend of intellectual stimulation and critical commentary. However, beyond substantive inputs, my committee provided an unfailing reservoir of emotional support. I am indebted to Anne Buttimer for her constant encouragement, and willingness to discriminate the wheat from the chaff; to Saul Cohen for having unwavering faith in me when others appeared to have lost theirs; and to Sandra Howell for bolstering a flagging morale on numerous occasions through her belief in the research. To Duane Knos, my advisor and friend, special thanks are due. His considerable time investment, his patience, and his commitment to helping me learn about the research process are greatly appreciated. But most of all I will remain forever indebted to him for providing the encouragement to continue on many occasions when I felt unequal to the task.

The completion of any research is facilitated by competent technical support. In this domain thanks are due to Margaret Jaquith who typed the original dissertation with speed and accuracy and who displayed considerable equanimity in putting up with a finicky writer's continual minor changes in material already typed. Similar fortitude and good humor was displayed by Shirley Lively of the West Virginia University Communications Department in supervising the typesetting of this revised manuscript. The publication of this book was made possible through grants received from the Department of Geology and Geography, and the College of Arts and Sciences at West Virginia University, and the West Virginia University Foundation, Inc.

Finally, it is a pleasure to acknowledge the help of my wife, Ruth. Her contribution extended far beyond the cartography. Over a three-year period she endured a spouse whose moods ranged from the depths of depression, through testy ill humor, to wearing exhilaration, in conjunction with the progress of the research. Without doubt she was the most tolerant of all.

GRAHAM D. ROWLES
Morgantown, West Virginia
November 6, 1977

PROLOGUE

In 1900, when my grandfather was a boy, there were slightly over three million Americans over sixty-five years of age (4.1 percent of the total population). By 1970, when he was enjoying his seventy-fourth year, there were over twenty million (9.9 percent of the population).[1] There has been a proliferation of studies concerned with this rapidly growing population.[2] Myriad sources document health problems, low incomes, job discrimination, social isolation, high propensity for mental illness, high suicide rates, and other indices of stress among the older population.[3]

The dismal statistics reinforce a public image. Attitudes toward older people are based on societal folklore involving an array of negative stereotypes. One such stereotype is that the elderly gradually become prisoners of space as physiological deterioration and environmental constraints necessitate physical, social, psychological and, by implication, spatial withdrawal. The specter of the impoverished, frightened, feeble elderly lady, eking out a barren existence, barricaded in a one-room attic apartment, haunts us all. Beyond uneasy concern for our own future, it arouses both sympathy and shame. In recent years society's collective conscience, prodded by the growing political voice of the elderly themselves, has prompted some action to "liberate" older people. Funding for housing, transportation, and health services, together with a host of new programs, appears to herald a new era of concern.

[1] U. S. Department of Health, Education and Welfare, *New Facts about Older Americans,* Washington: U. S. Government Printing Office, 1973.

[2] In this book an older person is operationally defined as a person over sixty-five years of age. This designation is an arbitrary choice. It follows a definition, first established by Bismarck in Germany in 1882, and later adopted as an administrative convention in the United States by the Social Security Act of 1935. Chronological age, however, is only one of many possible measures of aging. There are important differences among biological, psychological and social aging processes. Rates of aging in each of these domains vary considerably both for a single person and among individuals.

[3] A powerful synthesis of much of this literature is provided in Robert Butler's Pulitzer prize winning book, *Why Survive? Being Old in America,* New York: Harper & Row, 1975.

xv

Evaluating the impact of such programs it is often possible to provide indication of improved material well-being. But we are hampered in assessing the sensitivity of these efforts to many less obvious needs, by our superficial understanding of the older person's changing relationship with an environmental setting. This lack of sophistication is evident in the tendency to accept uncritically the notion of progressive spatial withdrawal with advancing age as a universal model, even though exceptions to this simple generalization abound. In part this naivete reflects the paucity of empirical research. However it also reveals a more fundamental problem. We lack insight, outside of an experimental context, into *experiential* aspects of the changing relationship between the older individual and his or her environmental setting. Yet there are many important questions in this domain. How does reduced mobility affect the way the proximal setting is experienced? What is the impact of the spatial separation of older people from their families, increasingly prevalent in contemporary society? How does familiarity, the legacy of seventy years of living, influence the individual's experiencing of the environment? Clearly, a simple model of progressive inexorable spatial withdrawal—growing old viewed as the reverse of growing up—is of limited value in considering such issues. It would be helpful if we could develop a more sophisticated interpretation of the older person's changing environmental experience. This book provides an initial step in this direction.

The study is an exploration. It probes, in detail, the person/environment relationships maintained by five older people (ranging from sixty-nine to eighty-three years of age), who share lengthy residence in a working-class urban neighborhood. In addition to investigating changing patterns of activity within the contemporary physical setting, the research explores many facets of their cognitive experience; including the manner in which the environmental setting is differentiated, emotional attachments to place, and vicarious participation in displaced environments of the imagination. The objective is to identify the principal dimensions of the participants' total involvement within the spaces and places of their lives, and to integrate these dimensions within a holistic conceptual framework which can serve as a model for interpreting the changing relationship between the older person and an environmental context.

The exploration is very much an interdisciplinary venture, straddling the boundaries of psychology, sociology, and anthropology, and deriving inspiration from literature in all three domains. However, I am by training a professional geographer. The perspective developed in the pages that follow is inevitably colored by my academic affiliation. It is thus appropriate to provide a brief explanation of why a geographer would concern himself with the matter of older people's environmental experience.

My research is located squarely within a *man/land* tradition of geographic inquiry.[4] Since the time of Hippocrates' *On Airs, Waters and Places,* scholars have interpreted patterns of activity in space within the framework of a man/environment interaction; sometimes emphasizing the dominating influence of the environment, and at other times stressing the importance of man's role within the equation. In recent decades growing interest in the study of environmental cognition has added a new dimension to the man/land theme.[5] In the emergence of this "perception" school within geography, the individual's cognition of the environment was at first viewed simply as a distorting perceptual filter mediating the man/environment relationship. More recently an increasing sophistication has resulted in the acknowledgment of the symbolic and emotional meanings of places as integral features of the total person/environment relationship. This more refined interpretation has been facilitated by growing appreciation of humanistically oriented approaches to inquiry. Moreover, in the quest for deeper insight, geographers are beginning to consider smaller groups and even individuals in their studies. Finally, there is increasing recognition of the desirability of considering person/environment relationships within a more holistic framework—incorporating social and psychological as well as physical dimensions of spatial experience.

The conception of person/environment interaction used in this book focuses on the notion of *geographical experience.* An attempt is made to acknowledge the multi-layered character of person/environment relationships—to incorporate forms of dialogue ranging from concrete expressions such as physical movement in space to the most

[4]William Pattison, "The Four Traditions of Geography," *Journal of Geography,* LXIII, 1964, pp. 211-216.

[5]For useful reviews and elaboration on themes developed in this paragraph, see, Thomas F. Saarinen, "Environmental Perception," in Ian R. Manners and Marvin Mikesell, eds., *Perspectives on Environment,* Washington: Association of American Geographers, 1974, pp. 252-289; Yi Fu Tuan, "Space and Place: Humanistic Perspective," in Christopher Board, Richard J. Chorley, Peter Haggett, and David R. Stoddart, eds., *Progress in Geography,* Volume 6, London: Edward Arnold, 1975, pp. 211-252; Torsten Hagerstrand, "What about People in Regional Science?" *Papers of the Regional Science Association,* XXIV, 1970, pp. 7-21; Anne Buttimer, "Grasping the Dynamism of Lifeworld", *Annals of the Association of American Geographers,* LXVI:2, 1976, pp. 277-292.

ethereal affective attachments to place—under a single rubric. In this context the word "geographical" means pertaining to space or place. "Experience," employing Webster's dictionary definition, is formally defined as "the state or result of being engaged in an activity or in affairs." This general definition embraces all the diverse modes through which a person knows or expresses his "being" within the world.[6] Using these definitions, the phrase "geographical experience" is derived as a designation for the totality of *the individual's involvement within the spaces and places of his life.*[7]

Although there are studies of older people's spatial behavior,[8] social scientists have not previously probed the *totality* of the older person's geographical experience. The research process is thus necessarily exploratory, with emphasis upon discovery rather than on the testing of *a priori* notions. In addition, the desire for sensitive insight into the intricacies of older people's geographical experience dictates that the methodology be personal and humanistic.

When I grow old I will experience what it is like to be elderly. Until then I can only speculate and seek to understand from the outside; for experiential awareness is personal and ultimately incommunicable. As Simone de Beauvoir notes: ". . .there is one form of experience that belongs only to those who are old—that of old age itself."[9] However, inasmuch as it is possible to attune my consciousness to align with the awareness of aged persons, to develop an intense form of empathy, I can derive more sensitive perspectives on their experience.

The quest for such intimacy was the essence of my research strategy. I developed strong interpersonal relationships with five older people. My conclusions result from almost two years' contact with

[6]Yi Fu Tuan, "Place: An Experiential Perspective," *The Geographical Review,* LXV:2, 1975, p. 151.

[7]Clearly, inasmuch as all experience occurs in space, this definition implies that all experience is geographical. This is, of course, the case. However, in this context, the designation geographical experience is limited to activity in which either the character of a location or the dimensions of space are integral components of the experience.

[8]See, for example, Stephen M. Golant, *The Residential Location and Spatial Behavior of the Elderly,* Chicago: University of Chicago, Department of Geography, Research Paper No. 143, 1972.

[9]Simone de Beauvoir, *The Coming of Age,* New York: Warner Books, 1973, p. 567.

these individuals. During this time, six months was devoted to intensive interaction, entailing at least weekly meetings with each person. The exchanges were unstructured. There was little data collection in the conventional sense. Rather, strong emphasis was placed on observation and on informal dialogue with the participants. Through this process I came to develop an appreciation of both the complexity and the many subtle nuances of their geographical experience.

In the chapters which follow, the research experience is described in a format which will be unfamiliar and perhaps disconcerting to some readers. Theoretical chapters are interspersed with a series of vignettes revealing the lifestyles and character of individual participants. Such presentation is more than mere capricious novelty. First, it attempts to convey a sense of the inductive research process. Second, I have chosen to include lengthy accounts of individuals in order to reveal the participants as creative human beings, adjusting in highly individualistic ways to changing personal and environmental circumstances. Finally, the inclusion of large quantities of data within the framework of vignettes provides a holistic context for individual observations, and facilitates interpretation by you, the reader, of the conclusions which were reached regarding the participants' geographical experience.

This prologue is more than the customary introduction. In addition to tracing the outlines of the problem explored in the pages that follow, the purpose is to solicit your indulgence. I ask you to share in the excitement of discovery, to travel the sometimes perplexing and often tortuous paths of a venture which turned out to be a challenging excursion into the unknown.

I'd like to start by telling you about my friend Stanislaw Linsky.

CHAPTER I

THE LAST DAYS OF STANISLAW LINSKY[1]

"Funny world, I tell you that. You struggle, struggle, work, work; and then when you get through work, you ready to die. That's the end of you."

The following obituary recently appeared in the Lanchester Evening Echo:

Service is Friday for Stanislaw Linsky, 68

Stanislaw Linsky, 68, of 23 Garrison St. died last night in the Mercy Hospital. Mr. Linsky was a steel roller at the Lanchester Works, National Steel Corp., where he worked 43 years before retiring in 1968. He was a member of the Century Rod and Gun Club in Wexford. He leaves a son, William P. Linsky of Lanchester; a daughter, Teresa C., wife of Peter D. Martin of Lanchester, and a sister, Mrs. Freda Doar of Bremen, Germany. Born in Poland, he was a son of Mr. and Mrs. James Linsky, and lived here 50 years. Funeral services will be held at 11 a.m. Friday in Collins Funeral Home, 623 High St. Malcolm Oatis will officiate. Burial will be in Lanchester County Cemetery, Elton. Calling hours at the funeral home are 2 to 4 and 7 to 9 p.m. tomorrow. The family requests flowers be omitted.

Thus, in a few terse unfeeling words, the life of a person I came to know was recorded for posterity. This narrative is not about "a steel roller at the Lanchester Works, National Steel Corp.," but about the human being with whom I shared several important months.

I was eagerly anticipating our first meeting. I had proceeded cautiously, slowly familiarizing myself with the locality and building contacts within the community. Several months had been spent reading about older people, developing a research strategy, and establishing a working relationship with case workers from the East Lanchester Neighborhood Center. A list of possible study participants had been compiled with the assistance of Louise, the case worker with whom I worked most closely, and some initial telephone contacts had been made. I

---

[1]In order to honor the confidentiality of some material in this study, the names of all individuals, locations, and identifying events have been changed.

1

considered myself prepared to embark upon the empirical phase of my research. Now I was about to meet my first study participant.

I remember feeling excited but apprehensive as we walked around to the rear of the three-decker building and climbed the wooden stairs to the second floor apartment. We knocked on the screen door. No response. For an instant I found myself hoping nobody was home. Then I heard someone shuffling towards the door. It opened. "Come in, come in," urged a gruff voice from the blackness within. Stepping quickly out of the November cold, we entered the warm darkness of his kitchen. And so it was that I entered Stan's life.

As Louise introduced us I felt embarrassingly self-conscious. Stan was a big gray-haired man. Later I learned he had once weighed over 350 lbs., and his size had earned him the nickname "Big Stan" in the neighborhood. Today he looked wasted, his flesh hanging loosely upon his giant frame. As I grasped his hand in greeting I noticed it was shaking and that hunched shoulders concealed his true height. In spite of this he projected an air of firmness, of stoic resignation, which served in a strange way to highlight a sense of suffering and yet at the same time imbue him with an aura of fierce dignity.

Louise started to explain the purpose of our visit. Stan listened politely but without enthusiasm. So I joined the conversation. I was a geographer trying to learn about older people's geographical experience. I wanted to find out what it was like to live in this neighborhood. I wanted to spend time in his company and to get to know him; to visit him regularly and perhaps travel around with him as he visited the places in his life. Displaying my tape recorder, I informed Stan of my wish to tape our conversations. Promising to explain everything I was doing, and to show him all I wrote, I guaranteed the confidentiality of the research. Stressing that we would have to meet many times I pointed out that we could use my car to travel anywhere he chose to go. I concluded by explaining that I very much needed his help in what to me was a very important project.

Not surprisingly, Stan seemed totally mystified by my amazing request. "What questions do you want to ask?" "I can't tell you nothing." "I've got nothing to hide." "Go ahead if you want to." "Whatever you like," were phrases I remember. In my notes on the meeting I recorded that:

> He seemed nonchalant about my presence, but indicated
> he was willing to be part of the study, although he expressed
> reservations about his ability to tell me anything I would want
> to know.

2

Her introductory role completed, Louise made some excuse about another appointment and departed. We were alone. We talked about Stan's background, about his childhood, the journey to the United States, a sojourn in New Britain, Connecticut, and his long association with the steel mill. Stan started to explain how difficult it was for him to get around these days. He stressed the restrictions imposed by his poor health, and confided that he did not have so many friends nor indulge in as many activities since he had retired. I began to relax. We chatted amicably although without commitment for half an hour.

Then Stan proposed we go out for a drink. This was working out even better than I had hoped. I was going to be shown one of his haunts on our very first meeting. Leaving the house we proceeded to Steve's Bar, a local bar known to Stan by its former name, The Imperial Cafe. Here we had a drink and talked some more. But after a few minutes Stan wanted to move on. I explained I had another appointment and arranged to meet him again on the following Wednesday morning. We walked together as far as Selena's bar, his next port of call, and there we parted company.

As I recall this first meeting, I remember Stan's obliging but less than eager cooperation. He must have thought, to use one of his own favorite terms, that I was "crazy." And in many ways I was. People just do not walk into other people's lives and blithely propose to become their bosom friends. Perhaps it was Stan's compassion which prevented him from summarily dismissing me. Perhaps it was his sense of humor. I am not sure. As for myself, I remember feeling elated as I walked home. It looked as though my research was going to work out really well. I had met a "typical" older resident of the Winchester Street area. All my stereotypes were reinforced. Only time would reveal my naivete.

Over the next six months I was able to piece together Stan's biography. Born in a small rural community in Poland, he was the youngest in a family of ten. He remembered ploughing with a team of two horses at the age of thirteen, and being involved in forestry work only a year later. "We worked up there. There was no loafing," he explained.

At the age of fifteen Stan came to America at the behest of his brother who had emigrated some years previously. He recalled arriving in New York on New Year's Day of 1920 and remaining an extra day on the boat because the immigration office was closed. Initially he lived in New Britain, Connecticut, working on a farm. Later he obtained employment at a local machine shop, but an argument with a new foreman terminated that job. So in 1925 Stan moved to Lanchester and started working for the National Steel Corporation. He was to remain

3

there until he retired. (One day Stan showed me the watch he was given
for forty years service. It was inscribed: "National Steel honors Stanislaw Linsky for forty years loyal and faithful service, January 1967."
The watch did not seem to have special significance for Stan. "I gave it
to my boy; he lost his," he commented.)

During his early years in Lanchester Stan lived in various rooming
and boarding houses on the north side of the city:

> I lived so many places, it's pitiful. While I was single, when
> I moved, I moved. When I didn't like some people who moved
> in, I moved out.

These were gay bachelor days. He would often talk of wild nights on
the town, dancing and carousing. It was around this time that he decided to forego the pleasures of driving:

> I decided I don't want to drive.
> Sometimes I was out, you know, with girls,
> and went out drinking. I was only eighteen.
> In the morning I had to look in the garage,
> see if the car was there!
> I wasn't for sure.
> So I says, "That's enough."
> I really realized I could kill somebody.

Towards the end of this period Stan met Josephine. In 1933 they
were married and moved in with Josephine's family on 29 Winchester
Street. Two children were born, Teresa in late 1933 and William in
1940. The marriage lasted almost nineteen years before, in Stan's
words:

> . . .she start fooling me;
> fouling me up, and I divorced her.

Apparently, Stan had objected to his wife's desire to work. This friction
was compounded by a trip to California during which she became "involved" with other men. Josephine subsequently married again. She
lived on High Street, not too far from Stan's apartment, together with
her eighty-year-old spouse who was almost totally deaf and partially
blind. On occasion Stan received a phone call from his former wife.

After the divorce, Stan remained at 29 Winchester Street for a
while but the arrangement did not work out:

> I was living with her sister.
> Hard time.
> I just had to move out of there.

4

And so in 1953 he moved two blocks away to Garrison Street where he was to spend the last twenty years of his life.

For much of this period, his son and daughter were living with him. Teresa experienced a brief unsuccessful marriage to a Korean War veteran. When he had returned, "they just didn't get along." She had recently remarried, and, at the time I knew Stan, was residing with her husband in a small house on the opposite side of the city. William, his son, was unmarried. He lived at home with Stan and worked in a machine shop only a few hundred yards from the house. In recent years he had become very strongly involved in the Jehovah's Witness movement.

Much of Stan's life in the years before his retirement was oriented around the routine of his work. He had had a large circle of friends, primarily drawn from the ranks of his co-workers. They would spend much of their free time drinking together in local bars. Stan's other passions were hunting and fishing. He described how he:

> . . .used to go hunting every day.
> Work from night, three to eleven.
> My brother (brother-in-law) come and meet me by the shop.
> Put my hunting clothes in a bag.
> Eleven o'clock.
> Change clothing.
> We go hunting.

In the past few years Stan had become increasingly limited by failing health. In 1964 he spent "four weeks and three days" in a New Hampshire hospital for treatment of his psoriasis. He described sojourns in four different Lanchester hospitals at various times. "The only one I wasn't in is University and Greenacres." In the summer of 1973, several months before I met him, it was discovered that Stan had cancer of the esophagus. Once more he found himself in the hospital, this time for a course of cobalt treatment. And so, at the time I met Stan, the world was very much closing in on him. A sense of resignation pervaded his whole being. He seemed to be waiting to die. "You got to kill time somehow," he would often remark.

As our relationship developed, meetings began to conform to a regular pattern. I would arrive at about nine in the morning and we would sit at his kitchen table with my tape recorder running.[2] At first I

---

[2]My tape recorder has a built-in microphone and so is relatively unobtrusive. After the first few sessions it was clear that Stan had become oblivious to the small black box on the edge of the table.

tried to follow a predetermined sequence of data-gathering tasks. One meeting was to provide a profile of his present-day activities, another was to reconstruct activity patterns of his past. There was to be a series of "mental map" drawing tasks involving sketching the neighborhood as perceived both in the present and in the past. And so on. It took little time to realize the futility of this strategy. Stan would not keep to the task at hand! Not wishing to be so constrained, he was disinterested in my formal agenda. Moreover, he had considerable difficulty with many of the tasks, especially those involving writing and drawing. Instead of the smooth logical progression of tasks I had envisaged, the sessions would become rambling conversations covering a wide range of topics. Some tasks were abandoned. Others had to be modified. Because of his difficulty with writing, a request that Stan maintain a diary of his activities for a one-week period became reconstituted as a series of visits extending over a week, during which I recorded his recollections of his sequence of activity during the previous days. Finally, one or two tasks which had not been planned but which emerged as practicable were introduced. In an effort to derive insights into his friendship network we spent some time going through his Christmas cards. On another occasion I produced a comprehensive listing of the older residents in the locality. I read this list to Stan and asked him to identify and describe his relationship with those people whose names he recognized.[3] However, in general it seemed that our most productive exchanges occurred during informal conversation.

After these conversations which generally lasted between one and two hours we would invariably go out. Sometimes we would walk but more often we would take the car. As I had promised, Stan was free to travel wherever he wished. I continually encouraged him to avail himself of the mobility afforded by my vehicle. The tape recorder accompanied us almost everywhere we traveled. Although much of the conversation recorded was aimless and disjointed, many insights into Stan's personality and experience were derived from comments he made as we drove from place to place.

We visited most of the bars Stan frequented. He would generally know the barman and a number of the patrons. Over the months we enjoyed much amicable banter with these familiar but often nameless people in his life. We would not tarry long at each establishment as Stan wanted to keep moving, ostensibly in order to prevent his legs from

[3]Unfortunately, Stan died before we were able to complete this task. Moreover, as he tended to know many people by sight rather than by name, this exercise did not provide a very sensitive indicator of his social network.

6

stiffening. However, he could have exercised simply by walking around in the bar. Perhaps there was a less direct interpretation. Barroom contacts seemed to provide a source of self reaffirmation, a sense of identity. But after initial greetings, the comforting pleasure of recognition, and a brief superficial exchange, there was not much to say. It would not be long before he felt it was time to move on.

A second focus of our travels was the shopping trip. As the weeks passed by this became something of a routine. On more than one occasion I was mistaken for one of Stan's relatives due to the frequency of our being seen together in the supermarket. First we would call in at Oreston's Tavern for a customary drink. After a few minutes we would cross the street and enter the supermarket. I would follow Stan around the store pushing a shopping cart in front of me. He would often pause to engage in conversation with other shoppers. Such fleeting interaction seemed to provide an important medium of social contact. He spent considerable time looking for the cheapest cuts of meat and those small enough for his meager needs. We would often shop at a small fruit store before making the one-mile journey home. On arriving back in the Winchester Street area I would usually drop Stan off at one of the local bars. I would take his packages home, place the perishable items in the refrigerator, and then travel back to the bar. I would return his key before joining him in a drink or proceeding on my way.

As we became closer friends, Stan began to take advantage more and more of my standing offer to transport him wherever he wished to travel. Gradually the diversity of our excursions increased (Table I.1). We traveled to bars Stan had not visited for some time. He was still remembered by many of the patrons at the York Hotel, but at the Boundary bar, and at the Rusty Harness things were different. Both establishments had come under new management, the decor was different and Stan no longer knew anyone. They had become alienating rather than friendly places. I also found myself making a wider range of service trips. On one occasion we delivered some clothes for cleaning and repair to a small run-down cleaners a mile from his home. We began to shop at supermarkets farther from the Winchester Street area. Another time we visited a factory shoe outlet where Stan purchased some new shoes. I remember him informing the assistant that he cared not whether they were black or brown. He purchased the first and only pair he tried on. On this same trip Stan decided to buy a new hat. This time he was far more careful in making his selection. On another occasion we traveled to the far side of the city to visit his son-in-law. As the weeks went by we ranged farther afield.

7

TABLE I.1
TRAVELS WITH STAN:
THE PATTERN OF AUTOMOBILE EXCURSIONS

Trip	Mileage	Maximum Distance From Home	Number of Stops	Nature of Stops
1.	5.9	1.9	5	3 bars***
				supermarket A
				fruit store
2.	4.2	1.1	4	2 bars
				supermarket A
				fruit store
3.	1.1	0.2	4	4 bars***
4.	3.0	0.9	4	2 bars
				supermarket A
				fruit store
5.	2.5	0.9	5	4 bars*
				supermarket A
6.	3.4	0.9	3	2 bars
				supermarket A
7.	2.8	0.9	5	4 bars
				supermarket A
8.	2.4	0.9	4	2 bars
				supermarket A
				fruit store
9.	2.3	0.9	5	3 bars
				supermarket A
				fruit store
10.	3.4	1.1	2	1 bar
				supermarket B
11.	7.3	2.5	4	3 bars*
				supermarket A
12.	4.2	1.1	5	3 bars
				supermarket A
				fruit store
13.	4.7	1.2	5	2 bars
				supermarket A
				fruit store
				New Lanchester Center
14.	3.8	1.3	4	1 bar
				factory shoe outlet
				supermarket C
				Azines Dep't Store
15.	3.4	1.3	5	2 bars*
				supermarket C
				Azines Dep't Store
				cleaners
16.	9.7	1.5	5	2 bars
				supermarket A
				cleaners
				Lanchester County cemetery
17.	5.5	1.1	4	2 bars
				supermarket B (closed)
				supermarket A
18.	15.7	4.4	6	3 bars
				supermarket A
				fruit store
				daughter's home

* = bar visited for first time.

8

There were two excursions I recall with particular vividness. Each took place after we had known each other for some time. I had been amazed when Stan admitted he had never been to the New Lanchester Center; it had been open for four years. The Center, a large indoor shopping mall, was the focal point of commercial activity in Lanchester. I suggested that we go there together on some occasion, and then abandoned the topic. Some weeks later, as we were returning from our regular shopping trip, Stan brought up the subject of the Lanchester Center once again:

Sometime I have to go with you to the Lanchester Center. I never saw it. . .I want to look it over.

Eagerly I encouraged him to make the trip with me. Allaying his fear of escalators, and assuring him there were not too many stairs, took some time. Finally, we resolved to go the following week. I was amused when he explained he would have to get dressed up for the trip. As I anticipated our excursion, I reflected that we had traveled back to many of the places of Stan's past but had yet to visit somewhere new. How would he react? Would he be filled with awe? Would he find the flood of color and the bustle of activity overpowering?

Stan was not impressed. He objected to the considerable walking distance from the parking garage to the main concourse of the shopping mall, and to the profusion of color and noise which made him feel uncomfortable. "How do you find what you want here?" he questioned. Most of all he was astounded by the high prices. "I can get these handkerchiefs for half the price at Azines," he observed. He was anxious to leave after a very short while. As we returned to the car I tried to summarize Stan's reaction. Bewilderment and disorientation were certainly involved but I sensed the essence was indifference. For a short time he had entered a different world and it was not to his liking.

I found a second excursion as bewildering as our visit to the Lanchester Center must have seemed for Stan. We had completed our shopping and were at a garage, where I had just been obliged to borrow some money from him in order to pay for my gas. Stan seemed restless. "Do you want to see my wife's father's grave?" he inquired. I was mystified. The cemetery was some distance away. On the way Stan talked a little about dying and informed me that he was to be buried in a cemetery in Elton. I remember feeling rather uneasy. I was not accustomed to this kind of morbid discussion. Eventually we arrived at the cemetery. Here I really began to feel uncomfortable. Stan could not find the grave! I drove all round the cemetery but he could not locate the spot. "I know it's near a water tap," he kept telling me. "It's somewhere on the right as you come in the gate." Finally, we parked and started searching among the gravestones. It was an eerie experience, and it took us more than half an hour to find the grave. When at last we found it, Stan did not wish to linger. He made a few endearing comments about his father-in-law and then suggested we leave. I was never able to fully understand the significance of this trip. Perhaps it had some relationship to a premonition of his own death just a few weeks away. Perhaps he simply wished to identify with the past for a while. Possibly he merely wanted to remain away from home a little longer.

As my friendship with Stan intensified, I became immersed ever more deeply in his world. A clear picture of his lifestyle began to emerge. His life seemed oriented to filling his days; to "killing time" as he phrased it. "So what do you do now?" I questioned, seeking a characterization of his activities since retirement, five years previously. "Hang around. Take a walk. Go in a barroom. Play a little cards." The response was resigned.

As his son rises at 4:00 a.m., Stan was often awakened early. Generally there would be desultory conversation before William left for work. Stan would rise between seven and eight, wash and shave, rub ointment onto his sore legs, and swallow his daily pill (cutting it into four pieces so that it would slide down his throat easily). Then he would enjoy a leisurely breakfast, usually orange juice followed by an egg and perhaps some oatmeal. As he slowly ate, Stan would start to read the paper. There was no rush. He could take his time. After studiously reading the paper, he would leave home and amble over to one of the bars he frequented. The morning would be spent moving from bar to bar. There were five comfortably within his walking range: Gervei's, Murphy's, The Half Moon, Steve's Bar, and Selena's. At each bar he would have a drink, or perhaps two, exchange a few pleasantries, share a joke with the barman or other patrons, and then move on.

At noon, or shortly afterwards he would return home. Sometimes he would heat up soup for his lunch. Then he would watch television and read his paper once more. Midway through the afternoon Stan would start preparing his son's dinner. William would arrive home shortly after four thirty and the two of them would usually eat together. By six, Stan would be ready to go out again. This time he would only travel as far as Gervei's or Steve's. A few more drinks. A game of cards, sometimes pitch, sometimes cribbage. Then home again. He would usually be home by eight. If William was at home they might talk for a while. More often he would watch television. Then, sometime between

10

nine and ten, he would go to bed. He would lie in bed listening to the radio before drifting off to sleep. This was the pattern of a normal day.

There were variations on this basic theme. If inclement weather precluded going out, he would spend the day drearily shuffling from his living room to the kitchen and back again, often pausing to gaze from his window at activity on the street below. On such days he would sometimes sit and watch the squirrels playfully scurrying up and down the solitary pine he could view from his seat by the kitchen table. Sometimes, when the weather was fair, instead of strolling between bars, Stan would feel well enough to take a longer walk. He might wander up Andrews Street in the direction of Killarney's Park, or perhaps up Winchester Street as far as Churchill Street, but he would never venture more than a few blocks from his home. On other occasions he would meet up with someone in one of the bars who would provide a ride to a bar somewhat farther afield. These occasional excursions served to keep him in partial, if infrequent, contact with some of his old haunts.

On Sunday Stan would remain home all morning. Usually he would listen to a religious service broadcast on a local radio station. On one evening each week he might go with his son to do the shopping and laundry. They would journey to Milton Square some two miles away. Stan would wait in the car while William attended to the chores. Then there would be an occasional, about once every two months, trip to the doctor's for a check on the state of his cancer. Finally, there were trips to the Pinewood Health Spa, about once every two weeks, to receive a steam bath and rubdown. On many occasions he told me how much he enjoyed these trips. Such variations, however, did little to relieve the monotony of his daily existence. Stan provided a fitting scenario of an alienated isolated older person. But the more I got to know him, the more I came to realize the superficiality of this image. I began to notice that Stan had carved a niche for himself within an informal but supportive social context.

One source of enrichment was provided by his friendship network. This had two overlapping components. There was a set of friends from his working days. Stan's co-workers had lived all over the city and in surrounding towns. He would still meet up with them on occasion and an exchange of pleasantries ensued. However, he never seemed to have had a close relationship with these people outside of a working context. As he explained when I asked about them:

What could I tell you?
I don't live with them.

11

I don't know their lives.
I only know they worked here.

There were exceptions. Igor Manichevic had been one of Stan's
closest friends. They had been drinking companions for many years.
Igor had lived not far from Stan's home. He had died just after Stan left
the hospital in the summer.

I thought he was good, and then I came out.
I read in the paper Tuesday, he's dead.
He died on Sunday.
On Friday or Thursday he visited me.
I didn't even go to the wake.
I wasn't good.

The loss of friends seemed a common occurrence. As he mused one
morning:

I meet a lot of people yet,
but most of them dying out.
Getting older and older.

As Stan had progressively lost contact with his co-workers, a second
component within his social network assumed greater importance. This
was a more locally based set of associations. There were the many local
residents with whom he would stop and talk during his walks. There
were the barmen. And then there were the patrons. Stan had a wide
circle of acquaintances here. Some were working men employed in local
factories. Others had been laid off or were not working because of
sickness. Many, like Stan, were in poor health. Others had simply fallen
on hard times, and still others possessed more dubious credentials.
There was the "Copper King," notorious for his less than spectacular
attempts to make money through fencing copper. There was "Ferret"
who spent a lot of his time at the Salvation Army detoxification unit
and much of the rest at Gervei's. There was "Peanut," "Red," and a
host of others. "You see, there's so many fellows I can't remember
them." These were the people with whom Stan would spend his time.
He talked about a frequent cribbage partner:

S. He's bad too. He's got a pacemaker.
    He's in very bad shape. He can't hear or nothing.
    Young fella, too.

G. Where does he come from?

S. I don't know. He lives some place. I don't know him.
He gets a pretty good pension but he ain't got a very
good life. He got landed with a pacemaker, you know.
He's got to have it often put in.
He's a sick man.
So he doesn't know what to do with himself either.

Stan could rarely remember the real names of these people and he did not know where most of them lived. Nonetheless, they were among his most important contacts. Providing a link with the ongoing world, they were constant reassurance that he was still alive. This circle of friends provided practical as well as social support. "Rick" at Gervei's would deposit Stan's Social Security check in the bank. "Steve" would purchase fish and other delicacies for him when he went shopping on River Street. "Dan," who worked in the trucking company across the street from Stan, would buy eggs for him and deliver them to his home. When the weather was poor, "Red" would often call on him to see how he was faring. Stan was not as isolated as it at first appeared.

Stan's family also influenced his lifestyle. Much of his activity at home involved preparing meals for his son and catering to his other needs.

When he comes home, the telephone rings all the time.
Can't even take a bath. I have to answer it.

According to Stan, this was a strong but somehow distant relationship. They would spend some time in conversation. In the evening William would often read the scriptures. The sessions would sometimes evolve into an argument as Stan did not have much sympathy with his son's religious convictions. Contact with his daughter was less immediate. Teresa often telephoned to check that all was well. She would visit several times a month and on occasion Stan would go with his son for a meal at her home.

In sum, Stan's daily life revealed a delicately balanced adjustment to the situation in which he found himself. His lifestyle reflected his limitations and yet, at the same time, allowed for maximum utilization of his perceived potential. This balance was reflected in his assessment of his physical potential:

If I sit still too long my legs get stiff now; you know they get
all numb and that. . .cramped up. When I walk too much they
get too tired.

Yet Stan's lifestyle also reflected his loneliness and search for companionship:

> I go out and watch television outside,
> talk to somebody.
> Get sick and tired of staying alone,
> you get tired.

His relationship with patrons of the bars fulfilled a need to be with people. He seemed in a constant battle with himself, a battle to remain involved in the process of living and to maintain vestiges of his former lifestyle. He grimly, resolutely, clung to life. Yet, underlying his actions there seemed to be a basic acknowledgment of the futility of it all: an aura of pathos permeating every sphere of his life. This emotional undercurrent found expression in Stan's negative attitudes and pessimistic interpretation of his role as an older person.

My primary concern in establishing a close relationship with Stan was to learn something about the quality of his geographical experience. As has been noted, geographical experience was defined as the individual's involvement within the spaces and places of his life. For Stan, geographical experience was a living collage; framed within the realities of a contemporary setting, colored by a rich reservoir of scenes from a lengthy biography and flavored with a sprinkling of fantasy.

On the simplest level, that of activity within the contemporary environment, Stan's geographical experience had become progressively impoverished. There were no more hunting and fishing trips.

> Two years ago I sold my gun,
> What I gonna keep it for?
> See if I keep gun, I get temptation,
> I take a license and go hunting.
> Then I can probably drop dead of heart trouble.

There were fewer excursions with friends; in fact, only occasional forays beyond the immediate locality. Failing health in conjunction with increasingly insistent environmental constraints seemed largely responsible for this. Even on an immediate scale, Stan moved around less and less. In the supermarkets he had become accustomed to asking assistants for the location of particular items in order to minimize the need to trudge up and down the aisles. It became difficult to negotiate long flights of stairs. He had become increasingly reluctant to travel on buses:

14

I ain't going to go in a bus.
I can't walk too good,
And I ain't going to go stumble,
and break my neck or something.

Even clambering onto a bar stool had become a laborious and painful chore. Clear illustration of the restrictive influence of the environment is provided by the crippling limitations which the presence of ice imposed:

When ice comes I don't go out.
Ah, when it is slippery, I'm afraid.
Snow I don't mind so bad, but ice, Oh.
Pretty hard for me to walk so much.
This is the worst weather for me.
This is where I don't trust myself,
because I'm afraid of falling down.
If I fall down, you know, I'm licked.

Because of such constraints, an increasing proportion of Stan's time was spent at home—his activity orbit had become progressively more localized.

Stan's movement pattern was closely linked to the way he conceptualized the environment. The local context was known as a series of resource nodes embedded within traversable space. His home, a few stores and the various bars, comprised the nodes. Stan seemed to possess detailed awareness of the most appropriate paths to reach each of them under varying weather conditions and at different times of day. He knew the hazardous intricacies of the sidewalk pavement, the cracks to avoid and the potholes to skirt. He knew where the traffic was busiest, and the prudent places to cross the street. In sum, he possessed a highly differentiated awareness of the local environment's structural identity. In zones farther from his home, the structural image did not seem so refined. Resource nodes (the doctor, the health spa, etc.) were set in space differentiated in terms of vehicular rather than ambulatory access.

For Stan, the contemporary environment was more than a structural arena. It was also heavily imbued with meanings, finely differentiated into a host of private and shared places. There were places in which to be alone, places to be with others. There were zones to be avoided:

The only place I wouldn't go is on High Street there, and
(those) places there they've got all the queers going.

There were homely places. Steve's Bar, Murphy's, the Half Moon, and
Gervei's all were pervaded with such a welcoming aura. There were also
places where he felt ill at ease, where he sensed alienation. The New
Lanchester Center was one such place. Locally, Selena's was too
"plush" and "ritzy." Laporte's and Moreno's were less than inviting for
different reasons. They were the province of a younger crowd:

I was too old for that, you know. They make noise, and I
don't want everything a lot of noise.

And so the environment could be subtly differentiated in terms of the
emotions it aroused.

It was impossible to understand Stan's feelings about his contemporary environment independently of his biography and his involvement in the ongoing history of the neighborhood. The significances
with which he imbued the spaces of his life were strongly colored by
events of the past. A fight at Murphy's, the fire at Selena's, the de-
structive encroachment of lumber yards into the district, the good fortune of the neighbor who won $50,000 on the State Lottery, the sad
death of the epileptic who lived down the street and whose body was
discovered three days after the fatal seizure—all were meshed within his
personal history and contributed to the differentiation of the contemporary setting. The meanings of spaces were the composite outcome of
association over many years. The environment was a living scrapbook,
harbinger of a million intimate secrets and nuances. Only as I grew
closer to Stan could I come to fully appreciate the tremendous importance of time within the totality of his geographical experience.

On one level it seemed as if Stan lived in environments of his
imagination, environments none the less real for their lack of contemporary concreteness. What to me was merely a dull and dusty street
could generate for Stan a veritable collage of landscapes. One day, soon
after our initial meeting, we were driving not far from Stan's home. We
started to ascend a long hill, flanked on one side by the grounds of a
local college and on the other by industrial wasteland. Stan began to
talk:

You know we used to go to work in the morning,
I counted up to twenty-one pheasants here sometimes. . .

It was a lovers' lane up here,
on this side farther down.

16

Sometimes we was going, eleven o'clock from work,
we used to count the number of cars,
sometimes up to twelve, eleven.
Then this motorcycle cop starts coming into them;
little by little he cleared them all out. . .

See this.
This used to be the foremen's club.
They used to come and eat here,
they had a cafeteria over there.
Where I worked they used to bring the junk,
what was left over there. . .make hash out of it, ha, ha!
The foremen got the best.

Not only did Stan reconstitute in his mind his contemporary context as it had been in the past, but also he was able to project himself into diverse geographical worlds far removed in space and time. Sometimes he would muse on the environments of his Polish childhood. On other occasions he would transport me to places where he had hunted. Once he shared a trip to New York State with me. The physical constriction which had come to dominate his life could not imprison his imagination. In this realm he remained free to travel at will.

Reflecting on my association with Stan I am aware that I gleaned insights which could only have arisen from a trusting relationship involving a degree of mutual dependency. Only by getting close could I share in his experience. What was Stan really like? What kind of relationship did we have?

Most significantly, I learned that Stan was not an "older person" but a unique individual. His experience was more a function of his character than his age. He seemed to sustain an uneasy balance within himself between counteracting tendencies. On one level he projected the typical profile of society's stereotype. He was physically frail, increasingly the servant rather than master of his environment. He moved around less, he avoided new environments, he withdrew from new experience, he rarely looked forward, and he was increasingly isolated from his former friends and colleagues. There was an overwhelming acceptance of the negative image of old age to which he had been conditioned. Suffering greatly from the loss of a highly valued work role, he felt wasted and superfluous. He had very much "bought in" to the negative societal image. Resigned to dull acceptance of the misery of old age, he appeared passive, morbid, pessimistic, without hope. He had given up on life and, as he expressed it on many occasions, was "waiting around to die."

17

Yet, probing more deeply exposed facets of Stan's experience revealing dimensions of more positive hue. There were glimpses of a person struggling masterfully to make the most of poor circumstances. He derived some solace from his family, and a sense of belonging and identity from relationships he forged at the local bars. His memories and involvement within environments of his past encouraged fond reminiscence (his eyes would light up, he would lean forward in his chair, and he seemed to come to life when we talked of Poland). There were positive elements in his personality which served to counteract negative tendencies toward withdrawal: a wistful dry humor, a sense of pride, an underlying resolve to continue. It was these dimensions of his being which encouraged his continued environmental involvement. And then there was our friendship.

I cannot tell what I meant to Stan. To suggest very much would be presumptuous, but I am convinced I was a positive influence. My involvement broadened his experience. In a modest way I was able to facilitate a temporary partial reversal of his constricting activity orbit. By transporting him to places of his past I was able to push back the clock. In some respects Stan came to depend on me both in concrete and less obvious ways. Somewhat to my surprise, he took me at my word with regard to my promise of the use of my car for his chores, so much so that there were times when I felt exploited! I also came to fill a social role and provide a degree of companionship. He would enjoy showing me off to his friends in the bars we visited. He seemed to look forward to my visits and at the end of each meeting would conclude with a questioning "See you next week?" I remember how hurt he seemed when I forgot a meeting on one occasion. But the benefits of our relationship were mutual. I received much more than I had anticipated; not only important insights into his geographical experience but also many of the benefits of a true friendship. There were ways in which I became dependent on Stan. When my work was going badly he would listen to my troubles. One day, during a period when I was feeling particularly anxious about falling behind my self-imposed schedule, he attempted to console me as we drove home from the supermarket. He did not say much. But there was a calmness and wisdom in his sympathetic ear and worldly comments on the enslaving power of time which served to reassure me. However, the relationship was not this simple. At various times I experienced emotions varying from love to disdain, from sympathy to anger. I'm sure Stan felt the same. It was a human relationship, not between an "elderly man" and a "student," but between two persons. Because of this closeness Stan's last few days were painful to both of us.

One day when I called Stan was not at home. On the previous week he had not seemed too well, but we had undertaken a particularly long excursion making many stops. By the time I let him off at his home he appeared exhausted. At the time his tiredness had not disconcerted me as much as his failure to conclude with his customary "See you next week?" On this cold morning I knocked several times but received no answer. It was not like Stan to miss a meeting. At first I had visions of him lying helpless on the floor inside. Dismissing such macabre notions I resolved to seek him out at the bars. Perhaps, after all, he had forgotten. I did not think so, but "Rick" or "Steve" might know where he was. I visited several of our usual haunts. Nobody knew where Stan was to be found. His absence had not been noted. I journeyed home and decided to call around the local hospitals. He was not in City Hospital, nor St. Paul's, nor Carlson Memorial. Eventually I located him. Five days previously he had been admitted to the Mercy Hospital.

During the next two weeks I visited Stan on several occasions. He was very ill. His throat was extremely painful and bronchial complications had set in. Often he could hardly speak. The oxygen mask he wore also impeded conversation. He looked pale and wasted, and as we spoke he would continually wipe large amounts of phlegm from his mouth with a tissue. We talked about our relationship. I tried to be optimistic, but he sensed he would never leave the hospital. "I'm not getting out of here," he informed me repeatedly, "I hope I go in my sleep."

I was embarrassed and I did not know what to say. I had not bargained for this kind of experience when my research started. As I sat by his bed, my mind would be a confusing welter of thoughts and emotions. Sometimes I experienced anger. "Damn it. You can't die now. I haven't finished my research." Immediately I would be overtaken by feelings of self revulsion. Did our friendship mean only this? Another time I toyed with the idea of continuing the research in the hospital—a tape from his bedside! For an instant it seemed an exciting possibility. Thus would I engage in the conflict between my human sensibilities and my scholarly purpose. The surprising outcome of this internal dialogue was that my work seemed less important, less significant in the overall scheme of things. I began to realize that my association with Stan had, in itself, been as important as the data I had derived from it.

19

Once I visited Stan in the early evening. I obtained permission from a nurse to enter his room. Stan turned slowly towards me. He grasped my hand. "You're a good boy," he croaked, smiling. It was a poignant moment and I was near to tears. He seemed so helpless. He motioned me towards a chair. As I reached for it, I caught sight of the charge nurse in the corridor. Out of sight from Stan she was beckoning me. I left the room. "You can't go in there," she informed me, "It is the wishes of the family." "But what about Stan's wishes?" I questioned, incredulously. "I'm sorry. It's their wishes." She was adamant. Despite my protestations I could not even return to say goodbye. It was suggested that I telephone Stan's son to seek reversal of the order, but when I called there was no one home. I left the hospital, angry at the indignity to which Stan was being subjected. His life was completely out of his hands, he no longer had any choices. The environment had become totally dominant.

I never saw Stan alive again for he died in the night.

CHAPTER II

AN EXPLORATION: PROBLEM, PROCESS AND CONTEXT

We shall not cease from exploration
And the end of all our exploring
Will be to arrive where we started
And know the place for the first time.

T. S. Eliot

In this chapter my experience with Stan is placed within the broader context of the exploration in which he was a participant. I explain my wish to undertake basic exploratory research into the older person's geographical experience. Essentially, the inquiry originated from questioning the substantive evidence for a pervasive public image of inexorable geographical constriction with advancing years (a demeaning image not only held by Stan but also internalized by many of his age peers). Preliminary assessment revealed the magnitude of our ignorance concerning older people's geographical experience and a need for careful empirical study in this domain. I describe the way in which this need became transformed into an intensive study of a small number of elderly persons living in an inner city environment, and outline the broad features of the research process as it subsequently evolved. The chapter concludes with an introduction to the environmental context in which my exploration was undertaken.

PROBLEM: THE OLDER PERSON'S GEOGRAPHICAL EXPERIENCE

The Closing Circle

What do we presume to know about the older person's relationship with his environmental setting? The essence of the prevailing image may be summarized as follows:

21

At infancy lifespace scarcely extends beyond the body, it expands as the senses develop and reaches a maximum in adulthood where it remains relatively stable and then with old age gradually diminishes until ultimately it stops at the body once again.[1]

The elderly gradually become prisoners of space. Certainly some individuals remain active and mobile into their advanced old age: the spritely centenarian is a celebrated social figure. However, he is primarily respected for the way in which he appears to differ from his peers and from the norm. For many older persons physiological decline, economic deprivation, and traumatizing effects of rapid societal change, herald physical, social and psychological withdrawal. This withdrawal, it is implied, is accompanied by progressive constriction of the individual's geographical lifespace, and associated intensification of attachment to the proximate environmental context.[2]

Spatial constriction can be viewed from two complementary perspectives: increasing *personal restriction,* and progressive *environmental constraint.* Physiological deterioration is associated with the aging process, although there is little consensus on the causes of this process and those components which are solely a function of aging.[3] There is a

[1]Leon A. Pastalan, "How the Elderly Negotiate their Environment," paper prepared for: Environment for the Aged: A Working Conference on Behavioral Research, Utilization and Environmental Policy, San Juan, Puerto Rico, December 1971, p. 2.

[2]The word "lifespace" is subject to a variety of interpretations, mainly deriving from the work of Lewin. (Kurt Lewin, *Field Theory in Social Science,* New York: Harper & Row, 1951.) Among gerontologists, Williams and Wirths cite Lewin directly, defining lifespace as "the entire set of phenomena constituting the world of actuality for a person or group of persons." (Richard H. Williams and Claudine Wirths, *Lives through the Years: Styles of Life and Successful Aging,* New York: Atherton, 1965, p. 3.) Birren employs a more limited definition: "The individual's lifespace is that part of the city he occupies physically, socially, and psychologically." (James E. Birren, "The Aged in Cities," *The Gerontologist,* IX:3, 1969, p. 164.) Clearly we are dealing with a very complex notion; each definition has its own nuances. In this study the term *geographical lifespace* is employed. The phrase borrows from Lewin, but the word geographical is added to emphasize spatial and locational dimensions of the totality of the experienced milieu. Such definition focuses the discussion on the geographical as distinct from the social or psychological world of the individual. Obviously there is a high degree of overlap among these domains. The differences merely reflect a difference of emphasis.

[3]For discussion of biological aspects of aging, see Diana S. Woodruff and James E. Birren, eds. *Aging: Scientific Perspectives and Social Issues,* New York: Van Nostrand, 1975.

tendency for stiffening of joints, calcification of ligaments, compression of the spinal column, loss of muscle power, decrease in cardiac output, circulatory system failure, reduction of lung capacity, impoverishment of tactile sensation, slowing of reflexes, decline of visual acuity and loss of hearing. Each of these changes contributes to reduced capability for successfully negotiating the physical environment. Consequently, a visit to the grocery store which was formerly a ten-minute stroll often becomes a tiring major excursion for aged limbs. One concomitant of these changes is increasing vulnerability to health problems, particularly chronic conditions. Arthritis, rheumatism, heart conditions, high blood pressure, and other degenerative conditions serve, if not to physically limit movement, to make it much more of a strain for the individual. Moreover, the constraints imposed by health problems may be accentuated by the individual's negative self-perception of his health.

Advancing age is also attended by selective changes in psychological capabilities: there is a rise in sensory thresholds; it becomes more difficult to absorb, organize, and evaluate environmental stimuli; and increases occur in reaction time.[4] These changes often engender feelings of incompetence and insecurity resulting from both actual and perceived loss of capability. Such insecurity may be supplemented by psychological withdrawal of the individual and increasing concern with inner feelings.[5] One outcome of this is that the older person often becomes reluctant to venture abroad.

Physiological and psychological changes resulting in the restriction of the individual's physical mobility are accentuated by crippling loss of role.[6] Old age is the first stage of life in which there is a systematic loss of status for an entire cohort. In addition, the present generation of older Americans is penalized by a lack of socialization to the fate of aging. There is no appropriate role model, for in the time of the present generation's forebears most people did not live for so long. Growing old thus involves an increasing sense of social alienation. There is no work

[4]James E. Birren, *The Psychology of Aging,* Englewood Cliffs, N. J.: Prentice Hall, 1964; Carl Eisdorfer and M. Powell Lawton, eds., *The Psychology of Adult Development and Aging,* Washington: American Psychological Association, 1973.

[5]Robert N. Butler, "The Life Review: An Interpretation of Reminiscence in the Aged," *Psychiatry,* XXVI, 1963, pp. 65-76; Bernice L. Neugarten and Associates, *Personality in Middle and Late Life,* New York: Atherton Press, 1964.

[6]Irving Rosow, *Socialization to Old Age,* Berkeley: University of California Press, 1974.

to go to, no socially defined responsibility: the individual stays at home.

Spatial withdrawal is only partially a function of increasing personal restriction. Such changes must be considered in tandem with powerful environmental constraints. The designed environment, particularly that of many urban areas, is not created for older people. Uneven paving and high curbs make walking hazardous, high steps on buses make boarding and alighting difficult. Design innovations which aid and even charm most younger populations can become architectural barriers. Stan's fear of escalators provides one illustration. More poignant perhaps is the reflection of an elderly London resident:

> What worries me is I can't get about like I used to, I'm always thinking about that. I only wish I could do more. I get so nervous too. They took me to one of those shows on ice. . . , I couldn't enjoy the show because I was thinking about them moving stairs. My son-in-law and my daughter had to carry me over them.[7]

The spacious layouts of many shopping malls, the wide expanses of modern transportation terminals, and, even today, in some environments the absence of places where a person might sit down and rest, provide further evidence of the manner in which the designed environment may be improved for the general population at the expense of the elderly.

There are other more subtle constraints. Often the timing of traffic lights is so rapid that older people are unable to cross without risking their lives. Small printing on signs, bus timetables, and bulletins, and the inadequate labeling of buildings serve to reduce the usefulness of these cues as aids to older people in negotiating the environment. Finally, the confusing color, complexity, and sheer speed of many contemporary environments may be seriously disorienting.

The poverty of many older persons restricts environmental participation by precluding options. The severity of this problem is indicated in the recent deluge of depressing statistics indicating, for example, that in 1971 some 4,300,000 older people (almost twenty two percent of the nation's elderly population) were existing on incomes below the

---

[7]Peter Townsend, *The Family Life of Old People*, London: Penguin Books, 1970, p. 283.

poverty thresholds for that year.[8] An outcome of this is that a trip to the movies, eating out, and non-service shopping expeditions, become major and infrequent investments. Not only is participation expensive but travel to resources becomes prohibitively costly.

Transportation has emerged as a major problem for the elderly.[9] Reduced income in association with failing health often precludes ownership of an automobile. A high percentage of older persons have no car. Although this loss can often be rationalized,[10] elderly persons anticipate and fear the consequent spatial stricture. Offers of rides from relatives and friends provide only limited compensation for the loss of discretion to travel at will. For many older persons fast disappearing public transportation provides the only means of travel. But public transportation is expensive. There are usually few buses between 9 a.m. and 4 p.m. when most elderly people wish to travel. In many areas the only option is to walk. Some elderly persons enjoy walking, but others are not well equipped for this form of travel.[11] Savings in terms of low cost are negated by other constraints. The demise of the corner store has meant that, for many, the local stroll to obtain groceries is no longer possible. There is considerable hazard, and consequent trepidation as a result of vulnerability to street crime. Finally, there is a very real fear instilled by traffic hazards which makes walking an unattractive option.[12]

[8]U.S. Department of Health, Education and Welfare, Office of Human Development, *New Facts about Older Americans,* Washington: U.S. Government Printing Office, June 1973.

[9]Edmund J. Cantilli and June L. Shmelzer, *Transportation and Aging: Selected Issues,* Washington: U.S. Government Printing Office, 1971; Stephen M. Golant, "Intra Urban Transportation Needs and Problems of the Elderly," in M. Powell Lawton, Robert J. Newcomer, and Thomas Byerts, eds., *Planning for an Aging Society,* Stroudsburg, Penn.: Dowden, Hutchinson & Ross, 1977.

[10]Frances M. Carp, "On Becoming an Exdriver: Prospect and Retrospect," *The Gerontologist,* XI:2, 1971, pp. 101-103.

[11]Frances M. Carp, "Walking as a Means of Transportation for Retired People," *The Gerontologist,* XI:2, 1971, pp. 104-111.

[12]While persons over 65 constitute only 10% of the population, they are victims in 25% of motor accidents involving pedestrians. Leslie Libow, "Older People's Medical and Physiological Characteristics: Some Implications for Transportation," in Edmund J. Cantilli and June L. Shmelzer, *op. cit.,* 1971, p. 16.

Compounding these environmental constraints is the tendency for older people to become spatially isolated from family and friends. The increasing locational separation of families, a feature of contemporary society, reduces the potential for face to face contact. There tends also to be a progressive spatial isolation from friends as the range of mobility becomes more limited and as age peers die:

> Residential settings of statistically "normal" age composition (approximating the distribution of the total society) may be excellent instruments for inadvertently alienating and demoralizing the elderly. With the attrition of ties with their family, friends, and other groups, the dispersal of their age peers in a normal neighborhood reduces the number of potential friends available to them. The field of eligibles is thin and scattered, and the isolating effects may be intensified by declines in health and physical mobility.[13]

Perhaps the most invidious external constraint upon the older person's environmental participation is a pervasive negative societal attitude. Society fears aging. The elderly are an ever present reminder of mortality. In a youth oriented society conditioned to revere the work ethic, the elderly are ignored, shunned, or at best, treated with indifference. In such a milieu the older person is made to feel useless and redundant. Little wonder that many older persons retreat into their homes and into their memories of a more equitable past.

Personal restriction and environmental constraints operate in conjunction. They support a model of progressive spatial constriction with advancing age. One result may be increased concern with the immediate environmental context. Older persons may gradually come to rely more and more upon local resources. There seems to be increased concern with remaining in secure and familiar surroundings. Some elderly persons do not venture out of a single room for days at a time. The immediate milieu, often limited to the home and the zone which can be viewed from the window, becomes the total contemporary world. As Gelwicks notes:

[13]Irving Rosow, *Social Integration of the Aged,* New York: Free Press, 1967, p. 39.

As more time is spent in the same setting, more and more psychological support is derived from objects near at hand. The proximal environment assumes an importance in the aged not often perceived by the mobile young adult.[14]

Beyond an obvious increased dependence upon local resources and proximate social contact, very little is known about the older person's relationship with the local milieu. Few studies have probed the phenomenological meaning of the proximal environment to the older person. We know only that it may assume great significance as evidenced by the fact that in many cases relocation can have pathological consequences.[15]

A dismal image of geographical lifespace constriction, expressed in terms of declining personal capabilities, increasing environmental constraint, and intensified affinity with the immediate setting, is rationalized by two of the most prominent early gerontological theories. The theory of disengagement focuses upon the social withdrawal of the older person. The theory interprets social and, by extension, spatial withdrawal and constriction of the lifespace as both a natural and desirable process. Disengagement is seen as:

> ...an inevitable process in which many of the relationships between a person and other members of society are severed, and those remaining are altered in quality.[16]

The process involves mutual withdrawal: withdrawal of the individual from society through increased introversion and preoccupation with self; and withdrawal of society from the individual through the sanctioning of social mechanisms such as retirement and widow status which provide society's permission to disengage. Successful aging is defined as a painless transition from a state of full involvement to one

[14]Louis E. Gelwicks, "Home Range and the Use of Space by an Aging Population," in Leon A. Pastalan and Daniel H. Carson, eds., *Spatial Behavior of Older People,* Ann Arbor, University of Michigan–Wayne State University, Institute of Gerontology, 1970, p. 157.

[15]C. Knight Aldrich and Ethel Mendkoff, "Relocation of the Aged and Disabled: A Mortality Study," *Journal of the American Geriatrics Society,* XI:3, 1963, pp. 185-194; Josephine M. Kasteler, Robert M. Gray, and Max L. Carruth, "Involuntary Relocation of the Elderly," *The Gerontologist,* VIII:4, 1968, pp. 276-279; Margaret Blenkner, "Environmental Change and the Aging Individual," *The Gerontologist,* VII:2, 1967, pp. 101-105.

[16]Elaine Cumming and William Henry, *Growing Old: The Process of Disengagement,* New York: Basic Books, 1961, p. 211.

of disengaged status. All elderly persons are claimed to experience such a withdrawal, although the transition is made more easily by some than by others.[17]

Disengagement theory is based upon a functional interpretation of society:

> The fundamental basis for the theory of disengagement is the mortality of man. All men must eventually die, and in order for society to outlive its individual members, some means must be found to carry out an orderly transition of power from older members to younger. This need is satisfied by the disengagement process. As one grows older the probability that death will occur increases, and at some point it no longer pays society to rely on the services of those who are about to die. For this reason it is profitable for society to phase out those whose possible contributions are outweighed by the possible disruption their deaths would cause to the smooth operation of society.[18]

In this notion lies a rationalization of the closing lifespace perspective on old age. The withdrawal of the older person is not only necessary for the functioning of society, it is also natural and desirable. What is convenient for society becomes translated into a metaphor for what is "natural" for an older person; a metaphor which leads inexorably to the notion of a closing *geographical* lifespace.

A second theoretical perspective, activity theory, also provides implicit support for the notion that as the individual ages, the geo-

---

[17]There have been many elaborations, evaluations, and criticisms of the theory. See, for example, Elaine Cumming, "New Thoughts on the Theory of Disengagement," in Robert Kastenbaum, ed., *New Thoughts on Old Age,* New York: Springer, 1964, pp. 3-18; George L. Maddox, "Disengagement Theory: A Critical Evaluation," *The Gerontologist,* IV:2, 1964, pp. 80-82; Arnold Rose, "A Current Theoretical Issue in Social Gerontology," *The Gerontologist,* IV:1, 1964, pp. 46-50; Gordon F. Streib, "Disengagement Theory in Sociocultural Perspective," *International Journal of Psychiatry,* VI:1, 1968, pp. 69-76; Robert J. Havighurst, Bernice L. Neugarten, and Sheldon S. Tobin, "Disengagement and Patterns of Aging," in Bernice L. Neugarten, ed., *Middle Age and Aging,* Chicago: University of Chicago Press, 1968, pp. 161-172; William E. Henry, "Engagement and Disengagement: Toward a Theory of Adult Development," in Robert Kastenbaum, ed., *Contributions to the Psycho-Biology of Aging,* New York: Springer, 1965, pp. 19-35.

[18]Robert C. Atchley, *The Social Forces in Later Life: An Introduction to Social Gerontology,* Belmont, California: Wadsworth, 1972, p. 31; see also, Gari Lesnoff-Caravaglia, "Senescence and Adolescence: Middle-Age Inventions," *The Gerontologist,* XIV:2, 1974, p. 98.

28

graphical lifespace becomes progressively more restricted.[19] Activity theory is based upon the widely held view that successful aging consists of staying active and being as much like a middle-aged person as possible. In contrast with disengagement theory, emphasis is upon a denial rather than acknowledgment of the inevitability of death. However, in this denial there is an *implicit* acknowledgment of a closing lifespace model. The theory has provided the rationale for many contemporary programs to aid the elderly to remain active. The very existence of these programs, focusing on the necessity to counteract restrictions imposed by aging, in turn, reinforces the internalization of a closing lifespace stereotype within public consciousness.

Thus far, a totally negative picture has been presented. How realistic is this view? To what extent is spatial restriction an inevitable attendant of advancing years? Are there no mitigating circumstances? Is there no contrary evidence? In recent years there has been an accumulation of evidence questioning the stereotype. Like Stan, many older persons are found to make extremely creative responses to changing circumstances. Moreover, as the diversity and individuality of the older population has become ever more apparent, perspectives have begun to emerge which support a more moderate and optimistic view of the capacities of the elderly to adapt and cope with both personal restriction and environmental constraints.

The Closing Circle?

Physiological decline and attendant health problems do not necessarily interfere with mobility. For many, such changes only become seriously restrictive in the final stages of life.

> Of the older population outside of institutions, 14 percent have no chronic conditions at all and 67 percent have one or more chronic conditions that do not interfere in any way with their mobility—that's a total of 81 percent with no limitation of mobility. Another 8 percent have some trouble getting around but can still manage on their own, sometimes using a

---

[19]Activity theory, also designated as the "implicit" theory of aging by Cumming and Henry (*op. cit.,* 1961, p. 13) has never been formally stated as a theoretical framework. Discussion of this perspective is included in Robert J. Havighurst, "Successful Aging," in Charles B. Vedder, ed., *Gerontology, a Book of Readings,* Springfield, Ill.: Charles Thomas, 1963, pp. 66-78.

mechanical aid. Another 6 percent need the help of another person to get around and only 5 percent are homebound.[20]

Prior to the final stages of life, controlled exercise and a carefully planned daily regimen can limit the impact of many physical decrements associated with physiological decline. For example, spectacles and hearing aids overcome, or at least ameliorate, the impact of sensory losses.

There is evidence of an ability to make subtle and creative psychological adjustments to reduced cognitive capabilities. As Nahemow and Lawton have summarized:

> Given the situation of a slower rate of information processing in old age, a counteracting adaptation may take place in an older person's capability to utilize increasingly larger chunks of information, thus, a heightened quality of "mature judgment" that is, the ability to discern the implications of environmental stimuli, or to plan.[21]

Also, appropriate therapy can lead to a reversal of psychological withdrawal.[22] Indeed, current social policy emphasis upon deinstitutionalization is symptomatic of increasing recognition of alternatives to progressive lifespace constriction.

There are even grounds for less pessimistic interpretation of the role loss which attends the aging process. As the years go by, an increasing proportion of older persons have parental role models to follow. There is a growing volume of documentation that:

> . . .older persons share the common denigrating beliefs about the elderly, *but only about others.* They exempt themselves personally from such invidious social judgments.[23]

[20]Herman Brotman, "The Elderly as a Significant Population Group," in Edmund J. Cantilli and June L. Shmelzer, *op. cit.,* 1971, p. 12.

[21]M. Powell Lawton and Lucille Nahemow, "Ecology and the Aging Process," in Carl Eisdorfer and M. Powell Lawton, *op. cit.,* 1973, p. 632.

[22]Leonard Gottesman and Elaine Brody, "Psycho-Social Intervention Programs within the Institutional Setting," in Sylvia Sherwood, ed. *Long Term Care: A Handbook for Researchers, Planners and Providers,* New York: Spectrum Publications, 1975, pp. 455-509; Carl Eisdorfer and Bernard A. Stotsky, "Intervention, Treatment and Rehabilitation of Psychiatric Disorders," in James E. Birren and K. Warner Schaie, *Handbook of the Psychology of Aging,* New York: Van Nostrand Reinhold, 1977, pp. 724-748.

[23]Irving Rosow, *op. cit.,* 1974, p. 88 (my emphasis); see also, National Council on the Aging, *Perspective on Aging,* IV:2, 1975.

In addition, a sense of social identity is provided by a growing age group consciousness, and a political activism not seen since the Townsend movement of the nineteen thirties. Evidence for the growing realization of Cicero's observation that "the crowning grace of old age is influence," is provided in the emergence of the "Gray Panthers," the National Council of Senior Citizens, The American Association for Retired Persons, The National Retired Teachers Association, The Retired Professional Action Group, and many other organizations. Such trends support the idea of emerging subcultures of the aged, defining appropriate roles for the older person.[24]

Environmental constraints can also be ameliorated or compensated for by older persons. Two sets of processes counteract restrictions imposed by the built environment. Public policies designed to create "barrier free" environments have meant that sloping inclines are replacing stairs, curbs are lowered or removed, glass panes are marked with color, and the built environment is being rendered more cognitively intelligible. Second, older persons are often extremely proficient at overcoming or circumventing environmental barriers. Many are adept at gauging the timing of traffic lights through the utilization of secondary cues such as other pedestrians or the sequence of traffic flow.[25] There is a propensity to avoid cognitively confusing environments, to patronize those which are more easily negotiable, and to more efficiently deploy available energy. Thus, Stan would ask a store assistant for directions rather than attempt to locate items by walking up and down the aisles of the supermarket.

With regard to economic constraints it is not as easy to question the notion of limitation. Increases in Social Security payments and the introduction of a Supplemental Social Security program have barely kept pace with inflation. However, the older individual can often reduce the confining impact of fiscal limitation. Efficient budgeting allows maximum return from limited income. Many older people pur-

---

[24] Arnold M. Rose, "The Subculture of the Aging: a Topic for Sociological Research," *The Gerontologist*, II:3, 1962, pp. 123-127; Arlie Russell Hochschild, *The Unexpected Community*, Englewood Cliffs, N.J.: Prentice Hall, 1973; Gordon N. Bultena and Vivian Wood, "The American Retirement Community: Bane or Blessing?" *Journal of Gerontology*, XXIV:2, 1969, pp. 209-217.

[25] For illustration of this point see Chapter VI.

chase major appliances (washing machine, television, refrigerator, etc.) before retirement in anticipation of their reduced income. Thus more capital is freed for investment in pursuits allowing the maintenance of an extensive geographical lifespace. And of course, not all older persons are poor. For some, limited income is not a constraining factor.[26]

Often, mobility restrictions are not as limiting as a dismal closing circle scenario would suggest. Golant, for example, found that a reduction in the number of trips taken by elderly Toronto residents was largely accounted for by cessation of journeys to work. For some subgroups he found that when the influence of journeys to work was discounted there was actually some increase in the number of daily trips.[27] Moreover, the proliferation of reduced fare programs, innovations such as the lowering of steps on buses, hydraulic lifts, and "kneeling" buses, and the advent of subsidized demand-actuated transportation, is tending to remove both physical and economic barriers to mobility. Again, creative individual adaptations are also important. Many older persons are able to secure rides from relatives and friends (Stan provides a good example here). Traffic hazards are avoided, and vulnerability to street crime is minimized by traveling at off-peak hours during the day when the risks are lower.

Community and church-based programs are an important influence in counteracting social isolation for many older people. Here too, there is evidence of an ability to make creative adjustments. Geographical separation from family may alter the medium of social contact but it does not seem to have strongly negative implications with regard to the strength of interpersonal commitment. Letters, phone calls, and other surrogate linkages provide substitutes for lack of physical proximity.[28]

---

[26]Recent evidence from a nationwide survey suggests that older people themselves do not feel as constrained by income limitation as the younger population considers them to be. Only 15% of a sample of elderly persons felt that "not enough money to live on" was a very serious problem for those over sixty-five, compared with 62% of the non-elderly who held this opinion. (National Council on the Aging, *op. cit.*, 1975, p. 6.)

[27]Stephen M. Golant, *The Residential Location and Spatial Behavior of the Elderly*, Chicago: University of Chicago, Department of Geography, Research Paper 143, 1972, pp. 142-156.

[28]As Osterreich has observed: "Geographical mobility is not disruptive of kin relations because the extended family legitimizes such moves and modern communication techniques have minimized the socially disruptive effects of geographical distance." It is further suggested that: "While geographical mobility may affect the types of interaction patterns found it should not result in a lessening of

Less direct environmental participation is also facilitated by the media. It is well documented that older people spend much time watching television or listening to the radio. Such pursuits serve to extend the older person's physical and social world by providing a substitute for direct social contact. Television personalities may extend the social world of the older person by providing a basis for "vicarious familism." Indeed, as Graney and Graney concluded from an investigation of older people's communications behavior:

> The evidence of the elderly person's ability to use alternative means to maintain social communication challenges the deterministic and fatalistic thesis that claims that the aged are powerless and become more so the older they become. In at least some areas of behavior, the elderly person's own interests continue to be pursued in aging, although in somewhat different ways than in younger years.[29]

Older persons are also able to make more active adjustments to potential isolation from friends and peers. The maintenance of informal and quasi-formal social networks in which the more able members provide both practical and social support to their less fortunate age peers is often an extremely important influence in facilitating spatial liberation.[30]

Finally, it can be argued that the constricting influence of societal rejection is becoming less potent. The recent outpouring of sympathy and concern over the poor image of aged status, in part an outcome of the rising political activism of the elderly themselves, has resulted in efforts to undermine the negative stereotype. As more positive attitudes toward the elderly, and an improved self-image among older people become more prevalent, the notion of inevitable spatial constriction is slowly becoming a less pervasive social myth.

In the same way that it is possible to question the universality of a closing geographical lifespace perspective, so too is it possible to furnish evidence which questions an image of obsessive prepossession with the

---

ideological and emotional commitment." Helgi Osterreich, "Geographical Mobility and Kinship: A Canadian Example," in Ralph Piddington, ed., *Kinship and Geographical Mobility,* Leiden, E. J. Brill, 1965, pp. 131-144.

[29]Marshall J. Graney and Edith E. Graney, "Communications Activity Substitutions in Aging," *Journal of Communication,* XXIV:4, 1974, pp. 88-96;

[30]For a good illustration of this process see Chapter III.

familiar and attachment to local place. For example, Frances Carp, in a study of older persons who moved voluntarily from community settings to public housing, found little reluctance to abandon both the former locale and the possessions of the past.[31] There is also a growing accumulation of evidence suggesting that it is more enforced rather than voluntary relocation which has pathological consequences. In some cases severance from familiar settings has been observed to result in improvement in various indices of well-being.[32]

Views implicitly questioning the inevitability of a closing geographical lifespace model can be subsumed under a rubric of "continuity" theories of aging. This perspective is based upon the idea that there is no single pattern of aging; that the process is the expression of successive personal adaptations to altered circumstances, and that autobiography is a key determinant of changes which transpire. The aging process reflects the continuation, with minimal adjustments, of lifelong patterns of physical, social, and by extension, geographical involvement. Much of the initial support for this perspective derived from more intensive consideration of materials assembled in the Kansas City Study of Adult Life which had spawned disengagement theory. Considerable variability in patterns of aging was found to relate to individual personality traits.[33] Analyses also revealed a typology of six distinctive aging lifestyle orientations: world of work, familism, living alone, couplehood, easing through life with minimal involvement, and living fully. For each of these styles it can be posited that distinctive patterns of geographical lifespace transition may be expected.

Translation of these essentially psychological and sociological findings into environmental terms seems to have been the basis for the emergence of what have variously been described as ecological, adapta-

---

[31]Frances M. Carp, *A Future for the Aged,* Austin: University of Texas Press, 1966, p. 89.

[32]M. Powell Lawton and Silvia Yaffe, "Mortality, Morbidity, and Voluntary Change of Residence by Older People," *Journal of the American Geriatrics Society,* XVIII:10, 1970, pp. 823-831; Ilene Wittels and Jack Botwinick, "Survival in Relocation," *Journal of Gerontology,* XXIX:4, 1974, pp. 440-443. For a useful review of the ongoing relocation debate, see M. Powell Lawton and Lucille Nahemow, *op. cit.,* 1973, pp. 640-645.

[33]Bernice L. Neugarten, Robert J. Havighurst, and Sheldon S. Tobin, "Personality and Patterns of Aging," in Bernice L. Neugarten, ed., *op. cit.,* 1968, pp. 173-177; Bernice L. Neugarten, "Personality Change in Late Life," in Carl Eisdorfer and M. Powell Lawton, *op. cit.,* 1973, pp. 311-335; Richard H. Williams and Claudine Wirths, *op. cit.,* 1965.

tional, or transactional perspectives upon the aging process.[34] Underlying all of these statements is the view that the individual seeks to maintain an equilibrium within the environmental context. This equilibrium expresses the reconciliation of available opportunities and resources with the individual's capability and motivation to utilize them: a reconciliation which varies with individual personality and autobiography. A model of progressive and inexorable geographical lifespace closure is only one of a number of possible relationships which an older person might live out within the environmental context.

New Directions

So we are left with a dilemma. What is the nature of an older person's geographical experience? Obviously there are both personal and environmental considerations suggesting a closing geographical lifespace. Equally, there is evidence that many older persons are able to overcome spatial constriction and that the process, at least until the final stages, is not universal. Why do some people seem to become increasingly attached to local place while others appear willing to abandon it with hardly a backward glance? Is there *anything* distinctive about the older person's geographical experience?

There is an even more important question here: do our fragmentary existing insights really help us in understanding the subtle complexity of Stan's involvement within the spaces and places of his life? Obviously, on the basis of present understanding, we cannot meaningfully interpret the pattern of daily trips to local bars, comprehend the significance of intimate knowledge of the neighborhood's physical and social structure, understand the influence of feelings for the various locations within his milieu, or appreciate the role of his awareness of place past. Yet all these themes are interwoven in the totality of Stan's geographical experience and are important clues to understanding his relationship with his contemporary environmental setting. Why can we not understand his experience on this level? There are several answers.

In most studies of older people's environmental experience there has been overemphasis on overt, easily observable dimensions of geographical experience, notably behavior. A serious gap exists because of

[34]M. Powell Lawton, "Ecology and Aging," in Leon A. Pastalan and Daniel H. Carson, *op. cit.,* 1970; Arthur N. Schwartz and Hans G. Proppe, "Toward Person/ Environment Transactional Research in Aging," *The Gerontologist,* X:3, Part I, 1970, pp. 328-332; M. Powell Lawton and Lucille Nahemow, *op. cit.,* 1973.

a failure to more than minimally probe the subjective worlds of older people. In particular, there is a paucity of empirical work focusing on the way in which older people orient themselves within their contemporary geographical milieu, and very little study of emotional identification with place. We have also noted that when the older person's arena of overt action becomes constricted, there may be important compensations in terms of surrogate experience. Yet there has been no study of the way in which such adjustments find expression within the totality of geographical experience. Finally, though few would deny that the older person remains free to venture far and wide in the realms of his musing, there has been minimal study of potentially liberating vicarious involvement in geographical settings. This is particularly surprising in the context of suggestions that vicarious experience assumes increased importance in the later years of life.[35]

A second consideration is closely related to this issue. Very little is known about the way in which *time*, as incorporated within personal biography, is integrated within geographical experience. The temporal dimension is obviously of utmost importance for persons who by definition have lived long enough to have accumulated a rich reservoir of experience and insight. Butler, in particular, has emphasized the significance of reminiscence in older people's lives.[36] But the significance of the passage of time and the import of personal history is even more poignantly expressed by an elderly hospital patient who observed that:

> Memories are an important part of living. Did you ever stop to consider the wonder of the mind that it's able to grab for things that happened long ago? That someone can say something like you just did, and the sound of your words makes my mind dance with memories. Not all of them good, mind you, but they come back as fast as you can imagine them. Very little of our lives goes away for good, you know, and that's a comfort to a man. That's a comfort to an old man who has known loneliness. No matter what they try to do to you, and in a hospital like this they can serve up all sorts of indignities. They can take away fifty feet of your intestines, but they can't take fifty seconds of your memory.[37]

[35]Bernice L. Neugarten and Associates, *op. cit.*, 1964; Erik H. Erikson, *Childhood and Society*, New York: Norton, 1963, pp. 268-269.

[36]Robert N. Butler, *op. cit.*, 1963.

[37]Thomas J. Cottle and Stephen L. Klineberg, *The Present of Things Future*, New York: Free Press, 1974, p. 49.

The remembering of events implies a remembrance of place. Remembering the time when the suburbs were green fields or forested slopes, older people may condition and interpret their lives with regard to environments displaced in both time and space. As a description of elderly residents in San Francisco recorded:

> They like to identify with a city of legends and character, and many old time residents are eager to recreate their personal involvement with the "old San Francisco" of the days before the 1906 earthquake and fire leveled it to the ground. They embroider their recollections with vaudeville headliners of the old Barbary Coast; robber barons transformed into society greats by their mansions on Nob Hill; the romantic history of old Chinatown and North Beach. These were the days when the gaslighter made his rounds at dusk and one could ride the cablecars for a nickel.[38]

We know almost nothing of the character and significance of the environments of fantasy and the way in which they are incorporated within the totality of the older person's geographical experience.

Both concern with subjective dimensions of older people's relationship to their geographical milieu and consideration of the role of time, are issues which can be embraced within a third, superordinate, area of consideration. At present there does not exist an integrated, consistent, and flexible theoretical framework for interpreting older people's geographical experience. Some studies focus on overt behavior patterns, others probe the psychological withdrawal of the older person from his environmental context. But none attempts to embrace the totality of the individual's relationship with his environment by considering the diverse modalities of geographical experience in concert. Such a framework must consider geographical experience in terms of the whole integrated person, recognizing the uniqueness of the individual, but at the same time seeking to express the interrelatedness among the various dimensions of geographical experience which is common to all individuals. By developing such a framework we can anticipate eventually being able to trace the older person's changing relationship with his geographical setting as it evolves through time.

The framework must focus specifically on the manner in which older people actually experience the environments of their lives. Notions about geographical experience should not rely almost ex-

[38]Margaret Clark and Barbara Anderson, *Culture and Aging: An Anthropological Study of Older Americans,* Springfield: Charles Thomas, 1967, p. 34.

clusively upon speculative inferences of parallelism culled from the inspection of findings in sociology and psychology: a procedure which has been necessary in much of the preceding review. Instead, insights should derive from empirical study of older people, focusing directly upon their geographical experience. At this point it is useful to reiterate our definition of geographical experience—the individual's involvement within the spaces and places of his life. By employing this broad definition, considerations ranging from contemporary movement in space to vicarious participation in environments of the imagination can be embraced under a common rubric. It becomes possible to seek a more sophisticated understanding of the older person's relationship with an environmental context.

The problem now becomes a methodological one of defining the most appropriate means for unveiling the intricacies of the older person's geographical experience. Clearly, lack of precedent, necessitates that the approach be exploratory. It also implies a need to become intimately immersed in the ongoing everyday experience of older people, in such a way that subtle often taken-for-granted aspects of their relationship with milieu may be revealed.

PROCESS: AN EXPLORATION

An approach aligning with a lengthy tradition of participant observation studies in exploratory social science research would seem to provide an appropriate strategy. However, in this study my quest was for something beyond merely the kind of insight which emerges from critical observation. I felt it was necessary to forsake the role of dispassionate observer and to become involved in—to become a part of—the experience I sought to understand. Such experiential involvement would, I felt, provide not only a more sensitive basis for developing insight, but would also foster a social climate in which it would be possible to share and discuss emerging notions with the participants.

Rejecting a traditional social scientific stance of remaining aloof and apart, I decided to establish viable open relationships with a number of older people. I hoped my friendship would, in each case, provide a reassuring context in which they might articulate a dimension of their lives—their geographical experience—they customarily took for granted. No attempt would be made to minimize the "contamination" of intervention within an existential situation. Rather, emphasis would be placed on the sharing of ideas through cooperative dialogue. I wanted the research interaction to become a mutually creative process. The result would be, I hoped, an experiential "text" in which were embedded the major dimensions of the older person's geographical experience. My job would then become one of translating from this text and distilling its essential geographical themes within a coherent conceptual framework—a framework doing minimal violence to the subtlety and integrity of the participants' lives.[39]

An in depth study involving a small number of participants provided the vehicle for my exploration. The original intention was to establish a close relationship with fifteen elderly persons from a common community context.[40] In the hope of reducing the complicating influence of residence in different environmental settings, it was decided to restrict participation to persons who had lived in a single neighborhood for at least thirty years. Apart from this, and a requirement that participants be at least sixty-five years of age, there were no constraints on the composition of the study population save for a willingness to engage in a lengthy research process.

Several procedures were used to establish contact with potential participants. A working relationship was developed with a caseworker (with special responsibility for elderly outreach) from the East Lanchester Neighborhood Center, possible participants were identified through consultation with local clergy, suggestions were solicited from leaders of local clubs and organizations for the elderly, and introductions were obtained through representatives of a Lanchester home care service agency. In addition, a short article describing the project and an advertisement soliciting participants was placed in the neighborhood newspaper. In spite of these efforts it proved extremely difficult to secure participants.

Once a potential participant was identified, the procedure was to visit the candidate in the company of a caseworker, or some other contact with whom the person in question was familiar. The project

---

[39]For a detailed account of this methodology, exploring its philosophical, operational, and ethical aspects, see, Graham D. Rowles, "Reflections on Experiential Fieldwork," in David Ley and Marwyn Samuels, eds., *Humanistic Perspectives in Social Geography,* Chicago: Maaroufa Press, 1978.

[40]I decided to study community residents for several reasons. There has been overemphasis upon the study of institutionalized and other partially captive populations, particularly those residing in high rise housing projects for the elderly. Second, the housing project population of Lanchester has been over surveyed. Finally, consideration of a community based population would facilitate some control of long term environmental variations which would have been more difficult to accomplish with a housing project population.

was explained, the considerable time investment involved emphasized, and the assistance of the older person requested. Little difficulty was encountered with this initial contact. Twenty-six individuals were visited. Two were subsequently discovered to have lived in the neighborhood for less than the requisite thirty years. All but three of the remainder expressed their willingness to be study participants. It was anticipated that, allowing for a reasonable drop out rate, this group would furnish between ten and fifteen study participants.

The problems emerged when a second visit was attempted. With distressing frequency, the older person would profess to be busy, unwell, or unable to see me for some apparently legitimate reason. Often I would be encouraged to call again after two or three days. However, each time I visited or telephoned there would be a new excuse. It seemed to make little difference whether I telephoned or visited in person; the response would be the same—polite avoidance. For example, in the case of Mrs. Auvergne a total of eleven contacts were made. Sometimes I would visit her apartment; on other occasions I would telephone to arrange an appointment. On several of these occasions I was able to arrange an appointment. But on each occasion she subsequently called me back to cancel our meeting. After three months of abortive attempts, I finally gave up. From an original group of twenty-one successful first contacts, seven individuals had to be discarded from consideration for this reason. I gained the impression that if all I had wanted was an hour to complete a questionnaire there would have been few problems in assembling a large sample of elderly respondents.

There were a variety of additional reasons for the progressive diminution of the study population. Mr. and Mrs. Rochefort, an independent defensive couple, became openly hostile. Timid Mrs. Boudreau, a fragile wistful lady living alone in a third floor apartment, suddenly decided to leave the district and live with her daughter. Mrs. Renoir, a proud self-reliant ninety-six year old, and her sixty-five year old son both entered nursing homes during the course of the field research. Finally, the withdrawal of the Armstrongs provides clear illustration of the sort of pragmatic difficulties which attended this phase of the study. We were introduced by a caseworker from a Lanchester home care service agency. David seemed friendly and cooperative but his invalid wife, whom I subsequently learned had not left the apartment for four years, seemed strangely threatened by my presence. I visited David several times and had begun to establish rapport. It quickly became apparent that his wife was the focus of his world; his life seemed oriented around the process of catering to her every whim. I suspect she began to see a weakening of her influence as an outcome of the relationship I was forging with her husband. On one occasion when I telephoned to arrange an appointment and was chatting amicably with David (he had already agreed to see me and was telling me how much he enjoyed my visits), his wife asked him to let her speak with me. "I'm afraid we're not very well. My sister died last week, and we're very upset. He doesn't want to see you," she explained. When I telephoned a week later, David informed me directly that his wife did not want him to see me. He was trapped from within.

Although a total of twenty-six older persons were considered in varying degrees of detail, an intensive relationship was developed with only seven participants—five individuals and one couple. The couple, a brother and sister who had lived together for over forty years, were excluded because a very different type of relationship was established with them than with the other participants. The continual presence of two participants resulted in difficulty in establishing a desired level of rapport with either. Thus five individuals eventually became the focus of the study.

The process of developing data also underwent considerable modification from my anticipations before entering the field. While the emphasis of the methodology was to be upon the informal establishment of interpersonal relationships, I felt that it would be difficult to develop rapport by commencing with meetings without any form of organized focus. Consequently, a number of formal tasks were developed both to derive "hard" data and minimize feelings of awkwardness which might arise during preliminary sessions. These tasks included content focused interviews, mental mapping exercises, maintaining a diary of daily activities for a one-week period, and structured autobiography sessions.

Such formality turned out to be not only philosophically at variance with the avowed intent but also impractical and unnecessary. The participants encountered considerable difficulty with some of the tasks. Attempts to develop a series of activity space profiles proved fruitless because of the participants' difficulty with remembering the times when they had changed their pattern of spatial behavior; mental mapping exercises proved of limited value because of difficulties with drawing and discomfort with an unfamiliar medium (a map) which, in addition, allowed representation of a limited range of phenomena; structured autobiography sessions were unsuccessful because of a tendency for the participants to direct the conversation away from the main theme (a three-hour exchange to probe facets of childhood experience of the environment would become limited to detailed explication

41

of a single incident); and diaries also were impracticable because of physical problems with writing and difficulties in articulating information in this medium. I found the participants were eager to communicate their experience more directly. In fact, the tasks seemed to get in the way and to increase social and psychological distancing. Fortunately, the folly of my aggressive approach was recognized early in the research and most of the formal *a priori* strategies were aborted before they could seriously jeopardize the development of dialogue or unduly prestructure the insights which emerged. Once these tasks were abandoned, relationships seemed to develop more naturally if less predictably.[41]

The research process thus developed on a more flexible basis. The primary focus of the data collection came to be open-ended conversation, tied in with activities of interest to the participants. Individual sessions might last for between one and five hours, depending on the nature of the exchange. The meetings would generally be extended if we decided to venture abroad—to walk around the neighborhood, go for a drive, visit a friend, or travel to a store. Sometimes the visit would be primarily social. We would play cards or watch television. There were times when we would become absorbed in autobiographical conversations. Often we drifted into grandiose discussions of the meaning of life. And there were sessions when we talked intensively on the significance of time, space and place. The terms would be different and much less abstract but the content was the same. In all the interactions, exchanges occurring in diverse contexts, I would be probing for insights into the nature of the individual's geographical experience. An effort was made to observe carefully, and to be open and receptive to subtle nuances in the individual's relationship with a total milieu.

On returning home, the remainder of the day, or in the case of evening sessions the following morning, would be spent on updating an informal diary, coding and transcribing taped material, or adding to my

[41]Once a basis of trust had been established, it became possible to engage in one or two more structured activities. But these tasks developed out of the relationships, invariably in collaboration with the participants. For example, it was found that a plotting of the locations from which Christmas cards had been received was helpful in developing some idea of the individual's social contact field. The derivation and plotting of family trees was a useful complementary procedure in defining the participant's social/spatial field. A third task which emerged during the research was the use of a listing of elderly neighborhood residents to aid in the identification of the individual's local peer group social network.

notes in the file I maintained on each participant. This process of intensive interaction was followed for a period of some six months.[42]

At the conclusion of this field phase I felt I had come to know a number of older persons with some intimacy. With each one I had established a distinctive relationship; a relationship facilitating some understanding of the individual's geographical experience. Many surprising insights had been derived. There was far less "hard" data than I had originally anticipated but I had accumulated reams of notes and many hours of taped material. The tapes proved invaluable in that it was possible to detect and corroborate themes or nuances I had been unable to sense in the original encounters. The tapes also furnished a constant reminder of the richness of the relationships I had forged.

Before introducing you to the remaining participants in the study, it is helpful to provide some introduction to the stage upon which the drama of their lives was played out, for, as will become apparent later, the scene is very much the complement of the players.

CONTEXT: "WINCHESTER STREET"

Figure II.1 depicts the area from which the participants were selected (hereafter referred to by its local "folk" designation, "Winchester Street"), and shows its location in relation to the Central Business District of Lanchester and a surrounding zone of transitional land use.[43] Until recently the entire locality fell within what Burgess termed the zone of workingmen's homes.[44] The area is geographically well defined, being bounded by railroad tracks and by a large cemetery. A pictorial map of Lanchester (dated 1878), unearthed during the course of the research, supports an inference that the locale has always functioned independently of surrounding areas. On this map there is a

---

[42] Although contact was not so frequent, relationships were maintained with surviving participants for a period of almost two years from the instigation of the field research.

[43] The Central Business District and zone in transition were defined according to criteria developed by Murphy and Vance and by Preston. Raymond E. Murphy and James E. Vance, "Delimiting the CBD," *Economic Geography*, XXX:3, 1954, pp. 189-222; Richard E. Preston, "A Study of the Land Use Structure of the Transition Zone Bordering the Central Business District," unpublished Ph.D. dissertation, Clark University, 1964.

[44] Ernest W. Burgess, "The Growth of the City: An Introduction to a Research Project," in Robert E. Park and Ernest W. Burgess, *The City*, Chicago: University of Chicago Press, 1925.

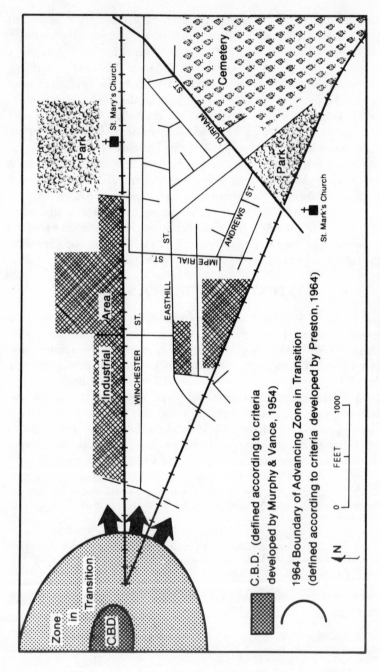

Figure II.1 Winchester Street

44

clustering of dwellings within the neighborhood which, at the time, was surrounded by expanses of open land. Extended conversations with neighborhood residents also testified to the social distinctiveness of Winchester Street throughout much of its history. This neighborhood was chosen for several reasons. One was its seeming geographical and social coherence. A second reason was the existence of a relatively large elderly population many of whom had lived in the area all their lives. Third, the locality was close to my residence, a factor facilitating intensive involvement and repeated visits to participants even during times of inclement weather. A brief outline of the neighborhood's history over the span of my study participants' lives provides a backdrop against which their experience may be evaluated.

A period of growth in the Winchester Street area parallels the rapid industrial expansion of Lanchester during the late nineteenth century which resulted in the decentralization of many thriving industrial concerns formerly located in the downtown area. The Restic Loom Works consolidated its operations in the Fulton plant on Imperial Street. According to local historians, after the merger this enterprise became "the largest corporation manufacturing looms in the United States." The Colton Brown Company relocated on Acton Street, and the Jefferson & Smith Tool Company also moved from the city core to a site a short distance from the neighborhood. In addition, Thomas's Carpet Mill (established in 1883) experienced growth and expansion on a site a few hundred yards from the area.

As the growth of these enterprises provided many new jobs, it was only to be expected that a large number of the immigrants who were flooding into the city at this time would choose to settle in the locality. Between 1890 and 1910 immigration accounted for a high percentage of an increase in Lanchester's population from 84,500 to 146,000. The Winchester Street area became the focus for three ethnic groups. By far the most numerous were the French Canadians. In fact, this group came to comprise a significant proportion of East Lanchester's immigrant population. The French Canadian population was very much oriented towards St. Mary's Church (built in 1904) as its sociocultural focus, and was the dominant group throughout the area with the exception of the southern section. The Irish, although far less in evidence, formed a second group. This population was fairly well dispersed throughout the area. Finally, the area became home for English families who had journeyed to the United States to take up employment at the Thomas Carpet Mills. The English, oriented towards the mill and served by St. Mark's Episcopal Church, were the dominant group in the southern portion of the neighborhood. Thus, the locality developed into a predominantly French Canadian working class residential neighborhood accommodating the employees of expanding industrial concerns.

The growth of the area was attended by the establishment of many local business enterprises. In the decades following the turn of the century thriving commercial foci emerged as Winchester Street became more and more an integrated self-sufficient neighborhood. Indeed, at this time the area was in many senses an urban village. As one of my elderly informants succinctly observed, "You could be born, educated, fed, clothed, treated in sickness, and buried, using services available within the neighborhood." Evidence of this phase remains even today in the survival of one or two neighborhood stores stemming from this period. Pasquel's, a hardware store established in 1913, remains in the family of the original owner. The City Market, purchased by the family of the present proprietor in 1915, still operates in the style of the complete neighborhood store. Further evidence is provided by the continuing existence of two of the most important unifying social institutions in the community; the French Social Club, and the British Men's Club which has existed on its present site since 1904.

The period between 1920 and 1950 seems to have been one of gradual diversification of commercial activity and a slow, almost imperceptible, erosion of strong local social identification. This was accompanied by gradual transition in the physical character of the neighborhood. In recent decades a process of residential stagnation and decay has accelerated as small warehousing concerns and lumber yards have become increasingly visible components of the local scene. Gradually, transition zone uses have infiltrated within the residential areas (Plates II.1-3). Large sections of the neighborhood have now been engulfed in the progressive outward encroachment of the zone in transition. This process has paralleled the outmigration, in fact the virtual flight, of the younger French and English residents from the area. But the elderly remain to live amidst decaying surroundings in mortal fear of a minority group renter population which has begun to move into the area.

The Winchester Street area is now very different from the vibrant community it was in the past. The major industrial concerns at the base of its early prosperity have fallen upon lean times. Today much of the district presents a dreary image. Many of the houses are dilapidated, stores are boarded, and the beer can and paper strewn streets provide a hollow echo of the neighborhood's former status (Plates II. 4-9). Only in ever diminishing residential enclaves in the areas farthest from the

46

Plate II.1 Infiltration of Transition Zone Uses

Plate II.2 Infiltration of Transition Zone Uses

Plate II.3 Infiltration of Transition Zone Uses

Plate II.4 Death of a Neighborhood

Plate II.5 Death of a Neighborhood

51

Plate II.6 Death of a Neighborhood

Plate II.7 Death of a Neighborhood

Plate II.8 Death of a Neighborhood

54

Plate II.9 Death of a Neighborhood

center of the city does the physical identity of the former neighborhood remain relatively untarnished.

The social character of the neighborhood has changed in concert
with this physical transition. Most of the English have either departed
or become assimilated. Other ethnic groups, particularly low income
Hispanic populations, are entering the neighborhood. Even the French
community has begun to lose something of its former social coherence.
Instead of an intimate, stable, family oriented social milieu, the quality
of the neighborhood has become more accurately typified by the neg-
lected children who play in the alleyways between crowded tenements, footloose teenagers who hang around the street corners, and
workers who stream onto the streets and into the local bars during the
midday lunch break. Much of the time the area projects an aura of
desolation and alienation. Fires of suspicious origin are not unusual,
vandalism is rampant, and violent incidents seem frequent (Plate II.10).
The neighborhood is in its death throes. It does not seem a hospitable
place in which to live.[45]

In spite of this, according to the 1970 United States census, Winchester Street was home for over 2,600 persons. Of these, 409 (15.5
percent) were over sixty-two years of age. A more detailed breakdown
derived from an alternative source, a listing of neighborhood residents
obtained from City Hall, suggested that in 1973 there were a total of
455 persons over sixty years of age and 348 over sixty-five. Many of
these older residents had lived in the Winchester Street area all their
lives. It was from this population that the study participants were
drawn.

In the following chapters an interpretation of the experience of the
four remaining primary characters is presented. Within these vignettes
lie major themes permeating the theoretical perspective developed in

[45]It is important to emphasize here that I have intentionally presented very
much the outsider's image of the Winchester Street area. As I became more
familiar with the setting, I began to attain a different perspective. Gans, in his
classic study, records a similar transition: "I developed a kind of selective perception, in which my eye focused only on those parts of the area that were
actually being used by people. Vacant buildings and boarded up stores were no
longer so visible, and the totally deserted alleys or streets were outside the set of
paths normally traversed, either by myself or by the West Enders. The dirt and
spilled-over garbage remained, but, since they were concentrated in street gutters
and empty lots, they were not really harmful to anyone and thus were not as
noticeable as during my initial observations." (Herbert J. Gans, *The Urban Vil-
lagers*, New York: Free Press, 1962, p. 12.) As will become increasingly apparent
as the discussion unfolds, the neighborhood to the participants, was in many ways
a totally different environment from the Winchester Street I have described.

2-ALARM FIRE SWEEPS BAR

The Fire Prevention Bureau is investigating a predawn fire which extensively damaged the main dining area of the Selena's Taverne, formerly Kathleen's Bar, at Imperial and Winchester streets. Fire officials said the lounge, owned by Mrs. Hazel Roney, may have been "fire-bombed."

Joseph W. Carre, district chief in charge, said the two-alarm fire, fueled by garbage, swept through the dining area and "went right through the roof." Engines 14, 2, 10, 13, 5 and 4 and ladder trucks 1 and 7 responded to the fire between 3 and 3:30 a.m., he said.

In February 1972, a man threw a Molotov cocktail at a waitress tending the bar, destroying the interior of the lounge.

School Vandalized

Vandals ransacked the Winchester Street School last night. Principal Peter Harmon said the vandals emptied files and tossed books around in almost all of the classrooms in one of the school's two buildings. A teachers' room, shown above, and the principal's office were ransacked. He said paint was poured on the basement floor, a fire extinguisher was emptied, and several windows were broken. He said the damage was discovered when the school custodian reported to work. He said he did not know the cost of the damage.

Winchester St. Man Foils Robbery Try

Two masked men, one apparently armed with a pistol, attempted to hold up a Winchester Street man shortly after 9 p.m. yesterday.

Daniel Flore of 194 Winchester St. told police the men walked into his apartment and attempted to rob him. The two men fled after Flore hit one of the assailants in the stomach, police said.

The two men, described as being five-foot-eight and five-foot-nine, were wearing nylon stockings over their faces, according to police.

Plate II.10 Crime and Vandalism

Winchester St. Man Shot; Woman Held

John McGill, 49, of 50 Winchester St. was shot in the chest and stomach in front of Murphy's Stroke of Luck, 38 Grant St., about 9:35 last night.

He was reported in fair condition at City Hospital, where he was taken by a police ambulance.

Fire Damages Three-Decker Rear Porches

A two-alarm fire of undetermined origin damaged the rear porches of a three-decker at 128 Winchester St. shortly before 2:30 this morning.

Fire officials said the blaze started on the first floor rear porch and spread to the upper porches before being extinguished.

The Fire Prevention Bureau is investigating.

Sentry St. Man Bound Over In Gun Death

Gerard Cambrid, 21, of 27 Sentry St. was bound over yesterday in District Court to the grand jury on a charge of murder in the Dec. 1 shooting death of Robert J. Brie, 19, of 74 Winchester St.

Brie died as a result of a shooting outside Steve's Bar, 242 Imperial St.

2 Men Beaten By Youth Gang

Henry Gould, 33, of 70 Imperial Ct., remains in satisfactory condition at City Hospital with injuries suffered when he and a companion were reported beaten by a gang of five youths about 6 p.m. yesterday.

Gould was admitted to the hospital with face and head cuts and multiple bruises. His companion, Charles Richmon, 42, suffered a cut mouth and bruises but did not require hospital treatment, police said.

Gould said he and Richmon left the VFW Post on Spenowbank Street and were jumped a short time later on Imperial Court.

Plate II.10 Crime and Vandalism *(continued)*

Chapter VII. In exploring this material, you the reader, are encouraged to ponder the geographical experience of each individual in terms of the speculative notions presented earlier in this chapter. In particular, note the variation in styles of relating to the Winchester Street environment: the diversity in patterns of spatial behavior; differences in the manner in which each person orients within the setting; the contrasting bases of emotional affiliation with place; the importance of contact beyond the confines of the area; and the role of imagination in coloring each individual's experience of the environment. Close scrutiny of these chapters will furnish a basis for critical appraisal of the ideas presented in Chapter VII, when an attempt is made to discern some order within the apparent chaos of the data. Let us first enter the fascinating world of Marie.

CHAPTER III

MARIE

"If I write my life, I'll tell you the truth here, that'll be the best story that you ever can tell."

The temperature was in the low twenties. It had snowed and the roads were icy as I drove down Winchester Street. Passing Marie's small brick house I noticed her tiny frame on the veranda. Wearing only a sleeveless dress she was busily cleaning her windows. One week previously she had been in bed with a severe cold. Marie was eighty-three years old.

The incident was typical of many during our association. She simply refused to acknowledge the constraints which age was supposed to impose on her life. On occasion her stubbornness verged on the foolhardy; it reflected the almost frenetic intensity with which she lived her life. As I came to know her better, admiration became tinged with a sense of pathos as I comprehended a certain desperation in her dynamism and obsessive good humor. Marie inhabited a rich and highly variegated world of her own making, a world in which she was righteous and successful. This world provided a strong protective cocoon. To admit alternative ways of constituting reality was threatening. It entailed acknowledging vulnerability and hence would furnish a basis for despair. Marie was steadfast: she would not yield. She would rage at the excesses of contemporary society, and cherish rosy images of an idealized past in order to reinforce her identity and avoid confronting an uncertain future. Superficially, she was totally unlike Stan.[1] She was active, involved, eloquent and outgoing. However, underneath, Marie was fighting the same battle, struggling for identity and the maintenance of continuity in an alienating world. She merely employed a different strategy.

Marie was tiny and slightly hunched. Sharp features, wispy gray hair, and a guttural French accent conveyed the impression of a wizened

---

[1]Although Marie and Stan lived within little more than a block of each other for almost fifty years, they inhabited totally different worlds. Stan did not know of Marie's existence, and Marie had never heard of Stan.

furtive character as she scuttled industriously about her home. However, after the cautious reserve of our initial meetings, I came to sense not just her ferret-like qualities but also a gentle warmth and an incredible desire to communicate something of the richness of her experience. She would greet me with an eager smile, seat me in her parlor amidst a profusion of dresses, skirts, coats, scraps of material and strips of yarn, and proceed to indoctrinate me with her view of the world. It was difficult to interject, and I felt vaguely guilty interrupting the steady stream of recollections and opinions. Time was precious. She had to tell all; and there was so much to tell. Leaving came to be quite a problem. There was always just one more story, one more anecdote, one more incident to recount. Sometimes I was obliged to beat a rather embarrassed undignified retreat. Four hours was all I could take.

Marie's willingness to talk made compiling a fairly detailed biography a simple task. Born in a small village in Quebec, she came to the United States when she was fourteen. Her family settled first in West Carlton, a small town eleven miles south of Lanchester. After three months the family moved to Chaugon a short distance away. It was here that Marie's father died. His death, only six months after arriving in America was a serious blow. It meant the children had to work. With scarcely concealed pride Marie remembers being obliged to lie about her age—"So much so that sometimes we didn't know our age," she laughed.

The children worked in local wool mills. The days were long and the pay was low. Marie remembers receiving 10 cents a week from her pay for spending money. The family would rise at six. They would be at the mill by seven. An hour for lunch at twelve, then another five hours' toil. Evenings were supper and prayers. Often there would be singing after supper, but by eight everyone would be in bed.

This lifestyle provided a model for the incessant industriousness which has characterized Marie's entire life. In spite of the hardships, she has positive memories of her childhood:

> When I was young I had a very happy life. . . ,and we had
> the best mother you can find.

Her resilience and optimistic personal attitude in large measure seem to derive from her mother's example of fortitude in the face of adversity.

Over the next few years the family moved on several occasions. There were sojourns in Milton, Drayton, and Eastridge (small mill towns to the south of Lanchester). In 1910 while she was living in Eastridge she met and married her husband, Robert. The couple moved in with his family. This was a very different milieu: there was no singing and little laughter in this family. Her mother-in-law was constantly morose, and a pall of depression had been cast over the family by the untimely death of a favorite son. Marie resolved at this time that her own home would always be "a place of happiness."

In 1917 Robert secured a job as a screw machine operator at Restic and Fulton. For a few months he commuted the twelve miles to Lanchester but with the onset of winter this became increasingly difficult. Thus the family (Marie had given birth to Suzanne, her first child, in 1912) moved to Lanchester and settled in the Winchester Street area.

At first they rented a small three-room apartment. The quarters were extremely cramped, and by the spring of 1918 they had resolved to move. Marie remembers meeting a man on Winchester Street and asking him if he knew of any apartments for rent. He could not help her, but pointed to a small brick house across the street and informed Marie that he thought the property was for sale. The family moved in on the 17th June, 1918. The house has been Marie's home ever since.

Fifty-seven years later Marie continues to lead a full life. Much has transpired in the intervening years. The home has witnessed birth and death, joy and tragedy. Two additional children were born, a son was killed during World War II, Robert died suddenly in 1951. Her home has been crowded and bustling with activity, filled with lodgers, children and laughter. Today it is quiet, a haven for memories. The dress-making business in which she once employed two seamstress companions been reduced to the point where Marie merely undertakes mending and alterations.[2] Now she lives alone.

Owning her own home, Marie pays only maintenance costs. She receives Social Security, and has pensions from insurance on her son who was killed and from her status as a Gold Star Mother. In addition, she receives income from her sewing. Conversations with other residents revealed she charges very little for her work and she privately confided that she employs a subjective scale of charges based on her own assessment of ability to pay. Much of the income from her sewing is given to a granddaughter whose husband abandoned her with six children. "So I help her," Marie remarked, matter of factly.

---

[2]"I used to have a woman working for me, Jennie Simone. She was living on Winchester down here. She worked for me at least fifteen years. And I had Mrs. Fuller, her mother-in-law, worked for me for seventeen or eighteen years. I used to run three machines down here. I used to have a big business down here. I was making all kinds of clothes. I was cutting; they were sewing."

> I'm happy to be able to earn money to give it away. I
> work like a poor devil, and I spend like a poor devil too. I
> don't keep no money.
>
> I don't need to sew. I have my pension and money to live,
> and I have my home. It's mine. I don't pay no tax, nothing,
> and I have more work than I need just repairing.

A small monthly stipend is also received for bookkeeping services to the Guilde, a small local insurance group. Marie has been secretary/treasurer since 1934. Finally, she occasionally receives a ten dollar bill from a local photographer for allowing him to photograph winter brides against the background of a pastoral mural which adorns the wall of her living room.

Marie seems unaware of the dire financial straits of many in her age group. We were discussing a Christmas dinner provided by the church for the elderly and disabled. It had been a marvelous affair according to Marie:

> We had a big banquet. There was over two hundred people,
> two hundred and forty people. We all bought our tickets. We
> had a caterer.

"What about the poorer older people?" I interjected. "The poor ones?" she replied incredulously.

> There's no more poor. Don't talk about poor people. The old
> people? There's no more of that. The old people now live
> better than the young ones because they have pensions. Every
> old people in the city of Lanchester have pensions. They have
> good pensions, too. So they all live like millionaires now,

she concluded emphatically. The more I pondered this exchange the more apparent it became that Marie's image of poverty differed from my own. It was framed in terms of conditions during the depression.

Marie continues to be extremely active.

> I'm still able to go out, my legs is good, my eyesight is good,
> and I'm bright enough.

Usually she rises between seven and eight. On most mornings she is to be found at her sewing machine. There are many customers and it was not unusual for me to find her crouched on the floor stitching a hem or advising on the appropriate button for a particular outfit. Working throughout the day and often far into the night, she rarely retires before midnight. There are frequent breaks: to eat, to read the paper, or to talk to a friend. In addition, the day is often punctuated by trips:

to play cards with friends, to visit a neighbor, to socialize at the French Social Club, or to attend a meeting of the church's "Golden Age Club" of which she is treasurer.

Minimal and grudging concessions have been made to age. She no longer walks "down city."

> I don't walk no more, because I can't go down street, that's too far, but I walk. I go down to my friend. I walk to play cards, not too far, but I walk.

Her granddaughter takes care of the weekly shopping and provides occasional rides to the doctor. However, Marie remains an active participant in her church, the French Social Club, and the "Golden Age Club." In addition, two mainstays of her life, her family and her friends, remain an integral part of her ongoing experience.

One of her surviving children resides in Florida, the other in Gary, Indiana. Two of her six grandchildren and their families still live in Lanchester. They make frequent visits and furnish both practical and social support. Moreover, her children generally visit at least once a year and Marie has made a number of trips to visit them. Geographical separation has not led to social isolation.

Over the many years of her residence in the neighborhood Marie has developed an extensive peer group friendship network (Figure III.1).

> I have friends. I have more friends than money, and I think this is the best thing you can have, and money is nothing to me. When you have friends you're always happy, and the more friends you have the better you can be. And I live for my friends, I live for my children, and this is what makes people happy.

Her friends not only provide a living link with the past but also more practical support, including rides to many of the functions she attends.

One particularly important component within Marie's friendship network is her "club." The "club" comprises a group of elderly ladies who meet regularly in one of the member's homes to play cards and socialize. Those who have cars arrange for transportation of the others to the monthly rendezvous. The "club" was born some twenty years ago. Originally it consisted of a dozen of Marie's closest friends. Over the years the group has developed into a self-perpetuating social unit. When a member dies, a replacement is invited to join the "club." Ostensibly this is to maintain a sufficient number to play cards. However, the effect is to maintain a viable cadre of older peers. As she has

65

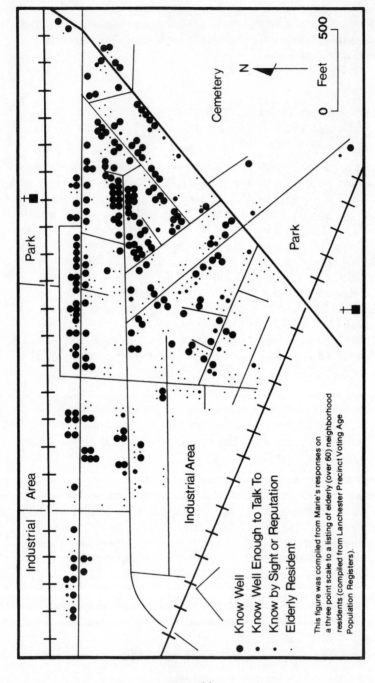

● Know Well

● Know Well Enough to Talk To

· Know by Sight or Reputation

· Elderly Resident

This figure was compiled from Marie's responses on
a three point scale to a listing of elderly (over 60) neighborhood
residents (compiled from Lanchester Precinct Voting Age
Population Registers).

Figure III.1 Marie's Winchester Street Peer Group Social Network

66

aged, Marie has seen this group become progressively younger as older peers are replaced by more youthful ones. As she wryly observed:

> All my friends are younger than me. The ones I had before, they died. They're all dead, the ones I used to be chums with, they're all dead.

An even closer relationship is maintained with one or two elderly neighbors. Mrs. Chanel, next door, has been an especially close friend.

> We've been neighbors so long we're like sisters, you know. We're not. If I need her, or she need me, I go, she comes.

Mrs. Delouvre is another close friend. She lives in an apartment two doors up the street and is a frequent visitor. Often she will spend the evening with Marie, and the two ladies frequently eat together or stay up late to view a talk show.

Overall, Marie has developed a lifestyle which is attuned to her capabilities and makes maximum use of local resources. I once asked if she had ever thought of leaving the area, of perhaps going to live with her daughter in Florida. She looked surprised:

> No, no, I won't leave! I have my own home, I don't pay no rent, I don't pay no tax. I'm living just fixing the house. I pay nothing. I'll be crazy to go down and live somewhere else. And I have my pension. I have everything to live for.

But there is another side to the story. The vehemence of her response provides a clue, an inkling that perhaps the superficially attractive image conceals more than it reveals. Probing a little deeper exposes a different Marie. To understand her geographical experience requires revealing her on this level.

Marie inhabits an alienating world. She feels both physically and psychologically threatened. One form of threat reinforces the other. Physical change in the neighborhood poses a basis for anxiety. In the area which can be monitored from her window, there have been few changes.

> There's not much change on Winchester. They changed a few houses.

But a few hundred yards away:

> There's nothing no more. It's terrible. Imperial Street now, it's not a street no more. It's a dump!

The zone of urban "blight" moves inexorably closer.

Social transition presents another dimension of threat. There is the "big building," a gray three-decker structure which Marie can view from her window. It symbolizes the first incursion of alien people into the social stability of the area adjacent to her home: a loss of "control" over the social character of the vicinity.

> In the old time the people didn't move like they move now. Now it's terrible, you know. We don't know our neighbors. They move, you know. Like the big building, you know, the people move. There's still two tenements empty there. They move out, they move in, they move out, so that we can't keep up. We don't know our neighbors no more—hardly no neighbors. On the big building I don't know nobody because they move in, out.

The change is frightening and somehow sinister:

> They had a lot of trouble down there. They didn't bother us, but the police was there all the time. The boy steal something; they were stealing bicycles. I don't know who they were. I didn't know them at all. They moved out last year. They were there only one year. The police came down and they took twenty-two bicycles.[3]

Social transition in the area has been paralleled by increased incidence of crime and vandalism. Marie is greatly threatened:

> We see some kids about thirteen or fourteen years old on the street at night. What business do they have out at night? They can steal your money right there. I'm afraid to go out at night. I don't go out at night no more.

Her level of fear is manifest in her response. Not only does she retire only when all her doors are locked and further secured with rope, but also she sleeps with a loaded pistol under her pillow and a wicked medieval-looking steel ball secured to her wrist by a leather thong. Intruder beware!

Physical and social threat is only one component of Marie's fear. It is symptomatic of something both more fundamental and more personally traumatizing; a perceived erosion of cultural identity and social responsibility, the demise of a moral order within which her values are deeply rooted.

---

[3]On subsequent retellings this figure was often inflated. On one occasion it transpired that forty-two bicycles had been stolen!

The life now is not the same thing as it used to be. We used to
know in the old time when I moved down here, everybody.
Everybody know us and we know everybody. They talk to us.
Now we don't know. There's no more friendship like there
used to be. We used to help each other's family. If one was in
trouble, if one was sick, we used to go down and help them,
but now there's no—the people ain't friendly.

For Marie to fully admit to her fears is, in a very threatening sense,
to acknowledge that the contemporary world is not for her, that she
has lived beyond her time. Her response is creative. It involves complex
and subtle denial processes coupled with an array of strategies facilitating self-maintenance. These may be summarized under a rubric of
two complementary reactions: aggressive insistence upon ability to survive in a contemporary society, and retreat into reverie and reinforcement of the meaning of a past order.

Insistence on ability to survive in the contemporary world is partly
a reaction to perceived threat and partly a denial of aging. There is also
an element of habit in the maintenance of an intensive level of activity.
I once asked her why she kept so busy all the time. "I have to," she
retorted, "I have to because, you know, I used to it anyhow." But
perhaps more significantly, Marie continues to work at her sewing business because her customers provide a continuing link with the community. They represent contemporary involvement.

Continuing involvement with clubs and societies, frequent excursions, and her periodic trips to Florida, to Quebec, or to Indiana, provide additional reassurance that she is not growing old. In fact, her
current pattern of activity represents a curious expansion of her activity
orbit. On the local scale she walks less and stays at home more than
formerly, but on a macro level she has ventured much farther afield in
recent years.

Underlying Marie's aggressive lifestyle is a defiant attitude. "If I
can't take care of myself, my place will be in a Home." She lives as
though there were something to prove to herself and to the world.

I don't go out with old people. All the ones I go out with, they
much younger than me, because I'm not one of those old
women. I can follow the young ones. What they can do, I can
do.

And even more poignantly:

When I change I can start to tell I am getting older [she was
laughing]. I'm just as happy. I can follow the rest. I can go any

69

place. I'm just as happy now because I don't believe in that. I go out. I went out to supper and danced the quadrille; went to the French Social Club and had a pot roast supper.

The second component of Marie's reaction is her retreat into reverie, a retreat complemented by condemnation of the present and negation of the future. Rejection of the present is typified by frequent tirades against contemporary society:

The young's not brought up no more. People don't know how to live with their children. It's not good for the children and it's not good for the parents.

You know, if everybody do like I do, there won't be no trouble in this world. But what makes trouble? Stealing, killing, like in Lanchester there. That's miserable now in Lanchester. We have fights. The paper's full of everything. If everybody was good there, there'd be peace on earth.

Negation of what is to come is reflected in gloomy prognosis of mankind's future. I asked her what she thought would happen:

That's the end of the world, the end of the world. That's coming, the end of the world. The end of the world.

The demise of makind is seen as an inevitable consequence of the total breakdown of the social order. So Marie derived solace and reinforcement in idealization of the past.

Sometimes she would dwell on particular incidents. There was the untimely death of the little girl next door, one of her closest friend's children.

The little girl was six years old. I remember that, because when she made her first communion I made a dress for her. And that girl was to make her first communion in June. And in the winter she went to school—Richmond Street. It was snowing and the water was high, and they make them walk, and when she come back she was all wet. She start to be sick, you know, cold and everything; and in three days she was dead.

Over the period we met, hundreds of such stories were recounted. On several occasions Marie told the story of a neighbor's apparently still-born child. The baby was not breathing and, according to Marie, even the attending physician had abandoned hope. Marie had picked up the infant and given him a sharp slap on the behind. The child let out a prolonged and lusty scream. "He have six children today," she would smile, basking in the glow of accomplishment. During each tale she

70

would muse on the significance of the past, once more involving herself in its ongoing process.

Many recollections were focused on her family. There was the fondly remembered and oft repeated account of her twenty-fifth wedding anniversary, when, wearing a new red velvet dress, she had danced with Robert until the small hours in the local hall down the street which had been rented for the occasion. She would recall accompanying her daughter on many of her dates. Suzanne had married Gaston, a young man living a few houses down the street. Marie would gleefully describe how Gaston jokingly protested that the couple had only been left alone together twice before they were married—and one of these occasions was when they went to see the priest! She even accompanied the couple on their honeymoon, acting as chauffeur for a tour of Quebec. One day she found her daughter crying in the cramped apartment across the street which she rented with Gaston. Marie was greatly troubled:

> So I said to my husband, "You know, Bob, we have only two living children. What are we going to do with the money we have? It belongs to the kids when we die. Why don't we buy them a house and put them in?" So we start to look for a house.

She would often talk of her son Paul who died during World War II. Originally he had been buried in England but Marie received notification that the body was to be transferred to Leavenworth, Kansas. She wrote to a high ranking military official:

> I took a letter and I write to him and I said, "This my son, and I forbid you to move him. Leave him in Norwich."

At the time she wrote Marie knew Paul had already been exhumed, but her letter provided the necessary leverage to arrange for burial in Arlington National Cemetery. She would tell how the family traveled to New York to meet the body. The military provided plush accommodation. After the funeral, a dinner was held in honor of the parents of the deceased. Marie was seated next to a general. She remembers his words:

> "I congratulate you, Mrs. Duvalle. This is where I found out about the heart of a mother. I have the letter you sent me. This will follow me in my coffin." He said, "This is where I learn what the mothers feel." And he took the letter out from his pocket and he showed it to me. And that general died only last year. And he never forgot me since my son died. I sent

71

him a card and he sent me a card. All at once I see that he sent me no card. So I sent my card the same, see. So the daughter told me that her father passed away, and he was buried with the letter. He had made on his will to follow the rules that he asked—that the letter he had in his pocket should follow him to his death.

Remembrance of time past implies involvement with place past. Each of the incidents described has a strong relationship with individual places; the house next door, a daughter's home, a revered burial place.[4] In themselves they assist in developing a collage of Marie's geographical world with its elaborate meshing of past and contemporary experience. She would also refer more directly to the places of her past. I was privileged to journey back with her.

Often she returned to her home of the nineteen twenties and early thirties. This home was a much safer abode. The family would frequently leave doors and windows open during the long hot summer nights:

> We used to leave all the toys of the kids outside, no but nothing was touched. The people used to go to bed at nights but now they don't go to bed. They roam the streets.

The home of reminiscence is a place of joy.

> I used to dance with them down here in the house. We had music. My daughter's a good piano player, and one plays the mandolin, the one who died in the war. And we used to dance. We used to take that table what I have down here, and we put it on the piazza, and we used to dance the quadrille down here in the kitchen. That was the happiness. And we danced all night—sometimes to one o'clock.

There were many variations on this theme. Each one reinforced her positive image. There was the story of the policeman ("On those days the police they were on foot."). One morning he had commented on the preceding evening's festivities:

> "You sounded like you were having a good time from all the shouting and screaming." I invite him in.

Her eyes would sparkle as she described him dancing around her kitchen. For Marie, the home of the past still lives.

---

[4]Marie has made several trips to Florida. On her way she always stops at Arlington to visit her son's grave.

There are times when she is able to do more than dream, when she can in effect turn back the clock. Once or twice each year the house is transformed for the visit of her son or daughter. I had grown accustomed to the total confusion of her home. There would be clothes, boxes, pieces of cloth, newspapers, and strands of thread strewn everywhere. A first reaction on entering Marie's home, probably that of a typical social worker, is one of sympathy. I remember my first visit. "Oh my, I guess she can't look after herself very well," I remember thinking. "She must be pretty confused." Just before Christmas I called one evening to find the home transformed. There was no material in sight, the carpets had been cleaned, and the mantles were covered with ornaments brought out of storage. The house had been recreated as it was. As she explained:

> Any time I have my son coming it's clean. It will be clean when my daughter comes in March. When she comes you won't see a piece of clothes anywhere.

In addition to her home, the neighborhood of a now passed era lives for Marie, not in a negative or pathological sense, but as a source of enrichment. She is aware of the physical and social deterioration all around her, but at the same time derives fulfillment in contemplation of a more auspicious past. She fondly reminisces on the time when Imperial Street was filled with stores and bustling with activity, the core of a vibrant neighborhood:

> We used to have a drug store on the corner, Lebouf's. It was a good one. Then it was a liquor store. There was a shoe store. And on Imperial Street there used to be a lot of stores. There was a grocery store there and jewelry shop and barber's shop, Pasquel's. And on the other side there was a fish market. Two grocery stores on the other side of the street. Mercier used to have a meat market at the back and grocery store at the front. They used to have a church there, was right on the corner of Easthill and Imperial. Father Deigneaux used to say the Mass down there. Imperial Street used to be lively. There was everything. There was a dance hall at the corner there, right at the bridge. The club there, we used to go down, dance the quadrille. All right at the back of the drug store. Now it's dead, there's nothing there.

More important than physical character, however, is remembered social identity. Marie revels in recollection of a sense of community she feels pervaded the area. The memories are sustained through interaction

73

with her friends, albeit younger friends than formerly, but nonetheless fellow denizens of past social place. Together they can maintain vestiges of an aura, a sense of belonging, the atmosphere of the neighborhood under a now failing social order. Together they can rue contemporary transition. For Marie then the social past is real. The place of the past is also the people and social order of the past.

Marie keeps many "scrapbooks" both literal and metaphorical. Her home is testimony to a full life. Each room, each corner, each artifact, is a cue to the past. The mural on the wall in the front room was painted by a cousin. He is now very sick and Marie has not seen him for many years. Several rooms were painstakingly decorated by a brother-in-law, "In the year the Lindbergh baby was kidnapped, 1934, I think." The home is a *memorial of birth and death.* Two sons were born here. Her husband died in the bedroom and his parents both spent their last days in the room off her parlor. The home is also a cue for joyful memories, it is the *seat of the family.* Her children played in the garden; they danced in the kitchen and cried in the parlor. Her daughter was married from here. It is a sacred place. At other times the house reveals itself as a *workplace.* It stores the memories of the comings and goings of a thriving business. A steady stream of customers. even today, serves as reminder. Finally, the home as scrapbook is a *symbol of achievement.* It expresses the success of her life and attainment of the American dream. After fifty-seven years the scrapbook is not easily abandoned.

Perhaps the most important place in Marie's home is the bureau in her parlor. Here she keeps her records, a concrete expression of her life. I remember vividly the day she first revealed these scrapbooks to me. We had been discussing her ancestors. Suddenly, she darted to her bureau. As she opened the doors a pile of scrapbooks, diaries, photographs, postcards, and newspaper cuttings slid to the floor. The cupboard was literally crammed with memories.[5] There was a scrapbook filled with newspaper accounts of a tornado, which, in 1953, had devastated a northern section of the city; a scrapbook on the old Lanchester, containing newspaper photographs of the downtown area; a scrapbook on the Kennedy assassination; scrapbooks on many presidents; a scrapbook on the Pope's visit to New York; a scrapbook on the World Trade Fair. There were also more personal documents. One

---

[5]In future meetings the mere mention of the bureau was enough to guarantee several hours of poring over the documentation of Marie's life and travels. It was fascinating, but towards the end of our association I was careful not to mention the bureau unless I had much time to spare.

scrapbook contained materials on the death of her son. It contained letters of condolence, correspondence concerning the transfer of his body to the United States, photographs of his crewmates, and other documentation of the tragic episode. An album of photographs of her ancestors and a booklet describing the community in Quebec where she was born, provide reminders of a happy childhood. Albums describing many of the trips she has taken record exciting excursions. They are stuffed with postcards, photographs, place mats, even hotel and restaurant bills. In some of the scrapbooks there is even a map of the route taken.

> Me, when I go on a trip, I mark the trip. It's all marked, I have
> the story.

The scrapbooks provide testimonial to a rich life and ever present cues for reliving the past.

Concern with maintaining an ongoing account of her life represents more than merely a basis for reflection. It involves a sense of continuity. Marie is concerned with providing a legacy for the future. Thus I experienced no difficulty in persuading her to keep a diary of one week's activities, and on my third visit I was presented with a lengthy unsolicited biography, painstakingly written in her scrawny script over the preceding week. Her son has encouraged her to maintain the scrapbooks, but despite Marie's eagerness to give them to him he has declined to accept them. A subtly cruel rebuff? She was more than willing to loan the precious materials to me: "You can borrow them. I don't care. What do I care?" I suspect she cares very much. Her willingness stems from a desire to be remembered, to ensure the continuity of her experience in a legacy to a world which might otherwise forget.

The neighborhood is also a repository of memories. It provides a rich assortment of cues. At first it appeared that Marie's affinity for the neighborhood was born of familiarity and identification with its physical character. Yet when I asked what she felt when a long familiar structure was torn down, I was surprised by her response:

> It doesn't bother me. If the people move, they take it down
> and they build something else.

I came to realize that it was the place as much as the structures which was important. What is important for Marie is the sense that something happened here; the mere existence of place is enough to kindle the memories. However, there are important symbolic exceptions. Across the street, on a tall slender pole listing toward the sidewalk, is a small black plaque with the gold star and gold lettering informing the world

75

that this is Duvalle Square. How often thoughts of a departed son who played on these streets must flit into consciousness? The neighborhood is a rich but subtle scrapbook.

Acknowledging the primacy of *time* is the essence of understanding Marie. Her contemporary life is an intricate meshing of past and present experience within the framework of a need to maintain a sense of identity and self worth. Stemming from this, her geographical experience involves the complex and creative juxtaposition of many geographical worlds in a manner which preserves the continuity and integrity of her life. A happy childhood and an active middle life, as well as her current environmental circumstances, are involved in her geographical experience. In the face of environmental deterioration and her own aging, she struggles masterfully, maintaining her integrity through a dual process of denying or rejecting the present and revering the past. Denying the grim realities of the present encourages continuing activity and, on the macro level, even some expansion of her lifespace. Affinity for the past is reflected in vicarious involvement in environments displaced in both space and time. Thus her autobiography becomes the key to understanding her geographical experience. Her contemporary geographical experience is more than merely her current activity pattern. It is the sum of the places she has been.

It might be argued that the creation and maintenance of such an individual reality is purely a naive defense mechanism, a withdrawal into a desperate coping mythology. But it is successful for Marie, and perhaps to some degree for each of us. Marie has achieved a viable if somewhat precarious wholeness in her world. Her strategy is not to be mocked. As she once remarked in a slightly different context, "Me, I'd be a good preacher to go out and preach the way to make people happy." Perhaps she could teach us how to accept the richness of our geographical past in deriving the most from our geographical present.

CHAPTER IV

RAYMOND[1]

"I may as well die happy as die crying."

Raymond was sixty-nine. He looked younger. Short, with gray-white hair, a round cherubic face, and something of a paunch, he conveyed an impression of health and well-being which seemed a perfect complement to his jovial good humor. He lived alone on the second floor of a three-decker building. This had been his home for almost forty years.

CELEBRATION

He was a long time answering the door this afternoon. "Come in. Come in. Come in. Come in." He looked bleary eyed but it was immediately apparent he was in an extremely jaunty mood. His daughter-in-law's parents had been visiting from New York.

> Gee'z. I didn't get to bed till four o'clock almost every morning. They were going to come down for New Year's but they were sick. They called up. "We're coming in. We're gonna celebrate. We're gonna await the arrival, so stay home Friday." So they came in by one o'clock Friday and we've been on the go ever since.

He motioned me to sit at the kitchen table where we customarily drank our beer.

> Well, Saturday, we waited all night Saturday. Nothing happened. Waited Sunday—until about four o'clock in the morning. Four o'clock in the morning, telephone rang. "It's a girl!"

Even now, three days later, he shook with excitement.

[1] Because Raymond was so articulate, this chapter is written with heavy emphasis on dialogue. Some paragraphs comprise a synthesis of material from several conversations. The words are Raymond's. Minor editorial changes, primarily omissions, have been made to preserve the flow of the conversation. The intention is to capture the essence of Raymond's experience, in his own words.

That's their first, their first grandchild, that the in-laws have. It's their first grandchild. I. . .what the hell, I've got six. I'm experienced, an old hand at it.

All my nephews and nieces kept on calling, Saturday and Sunday. "Anything happen? Did you hear anything?" So after the news, Sunday morning, early, I have to make the rounds, see. Call this one, call that one. "It's a girl. It's a girl." "Aaaaiee!" Everybody's been calling back. Sunday night at about eight o'clock the telephone rang. It's my daughter in Arkansas. "What the hell's going on?" she says, "I've been calling North Dakota all day and nothing happens."

Raymond was ecstatic. "So you wonder what a guy that's retired does when he's alone?" he gloated as he described the ensuing celebration:

> We went to New Hampshire Monday. We had to get juiced up a little bit because (laughs), see, liquor is cheaper in New Hampshire than it is here.

I imagined Raymond with his two elderly visitors careening along the expressway.

R. So we went up for a ride, and we went up to New Hampshire, up to Nashua. And we went to the liquor store and we stocked up, and we come back home. And course you got to—the bottles are nice, so you've got to empty them if you want to make something with the bottles! (laughter)

I'm telling you. They left this morning, this noontime. Jesus Christ, the glasses, the beer cans, I had everything there.

G. You've really been busy. So they had a ball?

R. Oh yes. They wanted to take me back. "Come on, come on, stay a month." But I knew if I go up there it's too wild and woolly. I'm. . .I can keep a certain pace. Now I got to stay and rest. I says, "Some other time. Give me a chance to rest up a little bit."

He moved over to the closet and opened it to reveal the remaining spoils of the New Hampshire excursion. There were three half gallons of whisky, bottles of brandy, gin and other more exotic drinks. I could visualize the revelers.

G. You've got a brewery here. I'm surprised you weren't unconscious. (laughter)

78

R. There's never a dull moment. Like I say, you know, you can be retired but you can still have a hell of a ball.

THE APARTMENT

It is Spartan.

I gave most of my stuff away. You know, I sent it away to my daughter, the fancy stuff, and antique stuff. And I just kept the dishes and stuff so that I could get by. But it gets dusty.

On the bureau there is now a clock which tells the time in Tokyo.

RETIREMENT LIVING

R. I get up when I feel like it. Like this morning, I woke up, half past six. That's too damned early to get up. I don't have to go to work. What the hell I get up? So I got a glass of water: I went back to bed, closed my eyes. Oh, just for a second. Quarter to ten it was when I look what time—quarter to ten! Them seconds are long. I get up and I make myself a cup of coffee, and I wash up a little bit. Read the paper. Then, when I get through reading the paper, I'll go downstairs, see the baby downstairs, talk with my niece downstairs and chew the fat. Other days I go cross the street. This is my niece across the street, she's Polish. She can give you a story about Poland. So I go over there and talk to her, and if she's down in the dumps, I pick a fight with her.

G. You pick a fight with her?

R. Yeah, I'll argue about something. Oh boy, it gets red hot; and I ain't going to agree with her. But after we get through arguing she forgot what she was down in the dumps for.
Then a lot of times we go out and pay the bills, telephone...

G. You don't mail them?

R. No. What the hell. I mean, you've got to get out. And once a month we go to the doctor's, get checked up. I take the old lady across the street; she didn't have no doctor. I fixed that.

G. Do you ever stay in the house for two or three days without going out?

R. Oh, if the weather is bad. I don't have to go out.

G. You don't feel cooped up?

R. Me? I can keep busy doing nothing. Oh I can keep busier than a one-armed paper hanger. People get bored because they don't know what to do with themselves. I'm always behind in what I have to do.

G. What difference does retiring make?

R. Well, retirement is like this. You wake up in the morning; look out of the window; there's a foot of snow; you see your poor neighbor going to work. You rap on the window and you say, "Why don't you jump back in bed?!" I see you up to your knees in snow. I don't have to go. I go back to bed. There's the difference. Before, you had to go because you always had a job. "I need the money. I can't lose a day's work because my kids got to go to school, or I've got to feed my family. I've got to do this, I've got to pay that bill." And then, you're regimented.

I said to Kerrigan, Monday night; we were watching the late, late, late movie. Two o'clock in the morning. "Hey Kerrigan," I says. "Now I want you to be honest with me." And we'd both got our glasses, we'd just refilled them. I says, "Now I'm gonna be honest with you. How about going to work? I'll get you a good job. You want it? I'll line you up with a good job." He looked at me serious. He says, "Are you kidding? I wouldn't go back to work for a million dollars. I'm so damn busy it's unbelievable." I says, "Back in nineteen forty, nineteen fifty, would you ever believe that you could retire and have a good time like this?" "No. Back then, if anybody had told me I'm gonna retire, I'd have called them a damn liar."

FAMILY, FRIENDS, AND FAITH

The family had owned property on both sides of the street. Raymond described moving from the opposite side into his present apartment:

When my mother-in-law died, I went to work one day. I had no intentions of moving. I come home at night and I'd already moved; I come home and I went up there and the place was empty. You come home from work, "Hey, you don't live here any more. You live over there!" It don't make no difference to me, you know. We were family.

80

I had a wife that was the nicest woman and as good as gold, very understanding, but she was sick for seventeen years. She had a heart condition, and then she had a tumor on the ear. I face my responsibilities. The last year she had cancer, see, and I kept her home. I took care of her, helped her, I nursed her, and everything else, till the end. And when I'd go shopping I'd take her with me, and I wouldn't go to the same store all the time—go to different stores to give her a chance to get out. I knew what I was facing. I knew what I had to do, make her as comfortable as I can. A good woman, good woman—and I got good kids.

He showed me his son's high school graduation yearbook.

Read what the people write about the boy. It'll give you an idea of what he was. He was manager of the basketball team, the four years he was there. That meant doing everything in them days. He had to see that the clothes was clean, see that everything was done. He had a couple of assistants organize everything.

Raymond's son, a captain in the Air Force, stationed in North Dakota, was transferred to Tokyo during the period of our meetings. I remember Raymond's proud reaction on first learning of the move:

Now I'm going to have a new place to visit. The boy has been transferred to Japan. He's got a big promotion. He's going with the Strategic Air Command, secret work.

Cecile, his daughter, had married and settled in Arkansas where she was raising a large family. He showed me an assortment of postcards and almost unintelligible scrawlings from his grandchildren.

R. I get letters from them. Now, how many kids would write to their grandfather? How many have kids like that? Now I get them from the three-year-old and the four-year-old, but it's all scribbled. But I have to write. I wrote six letters last night.

G. How do you feel about being here and them being out there?

R. Well, look, I feel this way. Life is life and you've got to let life go on. And I want them to live their life, not my life. You know what I mean. It isn't that I can't go there. The boy wanted me to go to Japan with him, and the daughter wanted me to go. But the roots are here. Nephews and nieces are just like my own. They come over.

81

In many ways Raymond is with his children. One day we were talking about his daughter in Arkansas. He reached for my notebook and, unsolicited, started to draw a map. As he sketched he talked, becoming progressively more deeply immersed in his reverie.

See, they got like a playground. There's a fence here and then, there's the other fence. Now in between this fence and that fence we make the garden on the right-hand side. And they dig sand and everything else. And there's swings, just like the playgrounds. On the other side is where *we—they* keep a horse and cow. And these are all trees. There's twenty-three trees, with grass. When I was there last year I made my garden. This was the garden. And then, right in here, I made four rows of potatoes, a hundred and fifty long. I planted the potatoes the second week of February, and Memorial Day we were eating our own potatoes. I had tomatoes, we had string beans and yellow wax beans, and peppers and corn, and pumpkins, and squash, different kinds of squashes, turnips, carrots, asparagus, watermelon, cantaloupe. I had them all.

Even after his return from Arkansas, Raymond continued to be involved with his garden.

R. I remember when, one time in June, no, in August, at the end of August, I was talking to my daughter on the phone, and I hear "Bang!" And she let out one yell, "Paul, get away from there." Paul had gone in the garden and come in with a watermelon. And you know he's only two, two and a half years old. And that's a long walk.

G. Have you ever had a garden in Lanchester here?

R. No. I don't have room.

Strong family ties also involve the extended family, his nephews and nieces and their children. This "family" remained in the Winchester Street area. Raymond basked in the recognition they lavished upon him. He would describe their response to the seven-month vacation he had taken in Arkansas and North Dakota:

Last year in December, the first week in December, I left here, and I come back the tenth of June. Now on three different occasions they call up. "When the hell you coming home?" And they're only nephews and nieces. The one across the street called up, and the first thing she said, she didn't say, "How are you?" "When the hell are you coming home?" Oh

82

they check up, they check up. The idea is, you know, you know they're thinking of you.

When I come back, my niece downstairs tell me. She said, about three weeks before I came home, "Saturday night we were having supper," and she says, "The girl was there and the boy was sitting on that side, sitting over there." And she was there, and her husband eating. Looked up. "What's the matter with you damn fools?" he says. The girl has tears coming down her eyes. And she looked at the eight-year-old and she looked at the boy again. "Sniff, sniff, sniff." And the mother said, "What's the matter? What are you crying for?" "We don't think Uncle Ray's coming back." And they were crying.

I'm lucky. I ain't one of those guys they hide, not to see or not to meet, you know. Some kids, they'll cross the street, but my nephew and nieces have all been brought up. Sometimes, sometimes it's embarrassing. I get in front of church. I'm all right when they kids, but they grown up now. Everybody see them, "Hello, Uncle Ray." And a big kiss, right in front of church. And everybody watching.

We're family, and yet I don't interfere with their way of life. I don't mean to say that the old people shouldn't fraternize with the young ones. But they shouldn't interfere with the life, the social life of the young ones. Oh, if they're invited, sure, but a lot of people, you know. . . If they do something that I think is wrong, that's up to them, they're old enough to know.

An extensive friendship network, built up over the years, formed a second underpinning of Raymond's world.

I forget their last names, a lot of them. It's a long time ago. Oh, hell, I know a lot of them. I know Mrs. Duvalle and, well, all those French people—you got the names. I visit them once in a while. I talk to them. I see them at church.

We were going through my listing of elderly neighborhood residents. He seemed to know everyone (Figure IV.1):

That was an old family that lived there. Her father died just a couple of years ago.

That's old. That Delbridge family lived on Imperial Street Court and then they moved to Andrews Street.

They run the television store, but they originally come from Richmond Street.

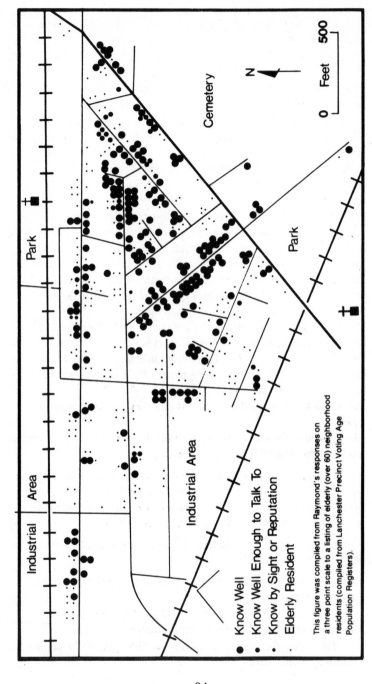

Figure IV.1 Raymond's Winchester Street Peer Group Social Network

Industrial Area

Industrial Area

Park

Cemetery

Park

N

0   Feet   500

- ● Know Well
- • Know Well Enough to Talk To
- ∙ Know by Sight or Reputation
- · Elderly Resident

This figure was compiled from Raymond's responses on
a three point scale to a listing of elderly (over 60) neighborhood
residents (compiled from Lanchester Precinct Voting Age
Population Registers).

They're related to her. They used to live on the corner of Tinterne and Cutler.

Her granddaughter married the wife's cousin.

That's not an original. They moved in in the last ten or fifteen years.

She was born in 1885. She still lives alone and maintains her own home.

I think it's her husband that used to be an ambassador. But she's the nicest woman. She's in her eighties. And she walks to Milton Square every day and carries bundles home.

Now she married a Lamarche.

That's the sister, she died last year.

That's an old family. They didn't always live on Sentry Street. They came from Winchester Street, that's originally, he's got brothers down there.

She's old and she'd be a good one to talk to. You can tell her I told you.

Her husband's just died. She used to play with my daughter.

Now that's an old, old, old, old, old, old family.

Now, Lambert, he used to live in a little cottage across the street there. Their son got killed in the war. And he was engaged to get married to, in Cutler Court, Lorion, a Lorion. She never got married—stayed an old maid.

There seemed to be a special relationship, a watchful reciprocity, among friends who lived within sight of Raymond's apartment. He spoke of Mrs. Boudreau, an eighty-six year old, living directly opposite:

I check on her; when she put her shades down. Almost every day I look out. If there's a light there and I see it, I'll wave. She knows that I'm home. And then if anything should happen she can grab the phone and call me.

And another neighbor:

He's one of the guys. Looks at the window. If he don't see me at night, sitting down on the couch watching television, he'll come bounding over here. See if anything's wrong.

Sometimes we would talk about especially close friends. David Armstrong was one. Raymond was in a playful mood:

R. When you see David Armstrong, I want you to pull something off for me. I'll tell you what. After you get talking to him you say, "Hey, I met the craziest creep there is in the city of Lanchester. I never saw a guy like him, he's so dumb, he's so crazy, he's so kooky." He'll ask you, "Who?" And you say, "I don't know, it's a guy, I don't know if you know him. He's the most insulting guy that you ever heard of. He lives over here, one of these streets. Cutler Street. No. No. I think, Sentry Street. A Frenchman by the name of Bachard, Bachard."

G. What will he say when I tell him that?

R. He'll tell you, "Oh, that son of a bitch!"

His eyes glistened with mischievous merriment.

Yes, David and I worked together down Cason and Sherrif. I've known David since he's been in Lanchester. And his wife, I knew his wife. Mrs. Armstrong, I knew her and her family before they got married, when they lived on Damian Street. That goes back to the twenties. They lived across the street from me.

His tone became more somber as he thought of his long-time friend, "He's slipping, he's slipping fast because he don't get out."

Then there were the Moynihans and the Connells.

We always were, we'll say, were close. You didn't have to go outside for activities. We'd have New Year's Night and New Year's Eve we'd spend together. I'd buy a bunch of lobsters, and we'd set the table up in the dining room there, and set the television to face the table, in the parlor, see. And we'd play cards until a certain time, till it was time to. . .we'll say half past nine. We'd get the supper going; cook the lobsters, stuff the lobsters, any way you want. And we'd set it on the table, and while we were eating we're watching the television, watching Time Square. And when midnight comes, "Whaaiiee!" We stayed close, you know. We visited, and we still visit.

Monsignor Richard was perhaps Raymond's closest friend.

I've got this priest that's a friend of mine, Monsignor Richard. I've been with him for close to thirty years, since he became a priest. On his days off, we take off. Sunday, I was over there, had dinner with him.

He explained how they would often discuss problems connected with the Monsignor's ministry:

You know, people call up with troubles. Well, if you haven't lived through it, you have a theory on it because you've learned it in a book. But books—most of the time you have to deviate a little from the books. So we get together. "How would you handle it?" "How would you handle it?" See. "Well, I'll tell you. Now if I was in this lady's place, and my kid had done this, and my kid had done that, I'd react this way and that way, and the different ways she did because I'm different than her. But seeing she reacts that way, you have to do this, and not this, which you'd do naturally this way because most of the people react that way." You know what I mean, we talked it over that way.

I hoped the Monsignor was not as mystified as I was!

Raymond possessed a deep unquestioning faith which provided a third supportive dimension in his life. Often he would talk about crises and describe how his faith had been justified. He would be out of work, his son would be ill, or there would be a difficult decision to be made. A favorite episode involved his daughter's discovery, soon after her wedding, that her husband had a drinking problem.

So she came home. "What am I going to do?" I said, "You married him, you go back." Sent her back. Got down on my hands and knees; asked the good Lord to take care of it. She went back. He hasn't smoked a cigarette, and hasn't taken one drop of liquor ever since. That goes way back in 1952. Crises are gonna arise that was more than I could handle. I couldn't handle it. It had to be somebody intervened and changed the mind of. . .or make them veer off and wide.

I always say there's a reason for everything, there's a reason for everything. If you sit down and follow things and watch things develop, in time you'll see the reason. You never see the reason right away. Whatever you thought was a catastrophe, was gonna happen to you. Look back and see—gee, there was a reason for that. But people don't stop and think. They won't give credit where credit is due. What do they know? Nothing. These, the same people that go to church? For cripes sake, they don't even believe in it. They're one foot in the church and one foot out. The reason you go to church, it's supposed to be, to meditate in the temple of God. And when you're in there you talk to him, tell him your troubles and have faith in him, and he'll come down.

Now I don't know, you don't know, nobody knows, whether there is a God or not. We didn't see him. Right? Now we base our way of living on our religious life, our faith. Now our faith, if we call ourselves Catholics, Episcopalians, Methodists, anything you want, you can call yourself that. But are you really a Catholic, really an Episcopalian, or really a Methodist, unless you participate and learn your religion? You know what I mean? Oh, you can quote the Bible and everything. What the hell good is quoting the Bible if you're ready to shoot your neighbor?

When I read the Bible. . . Now there'll be people are gonna deny the Bible, this and that. Who got the Bible together, who wrote it, and everything else. Any book you write, there's got to be some truth in it, in a part of it, no matter what story you're telling. Guy's got to have something to write that story about which has happened. Now I can write a story, we're gonna say. You got to write something about drinking a can of beer. You over there, me over here. I'll write a story and you'll write a story. And when they read that thing it'll be two different things. But the basis of it is the same thing. They all wind up drinking the can of beer. Right? So the Bible always refers to the temple. The Lord sent Peter to build a new church. Why did he say that? What do you need a temple for? It's to gather the people to pray together.

I asked about his association with the local church, Saint Mary's, a massive imposing structure overlooking the neighborhood, and a focal point for the French community:

R. Well, St. Mary's was important to me because, like I says, I put a lot of work in there.[2] All my nephews, nieces, went there. The kids went to school, graduated from there. My daughter was in there, my boy was in there. It's, it's in your life. Instead of going to the night clubs, and sit down and watch somebody or other, you're having a good time with your own crowd.

G. What makes people your own crowd? Living in the same area? Being French? Being in the same church?

[2]At one time Raymond assisted with coaching the basketball team. He had also organized many excursions and, in association with another parishioner, had run a boy scout troop affiliated with the church. Like many of his elderly French peers, he had given generously to the church.

R. No, no, no, no. Your own age that you meet. Now you don't know nobody, and you go down to the church. All of a sudden you'll come in, to a supper or something, and you look around. And you ain't gonna go sit down with old, old, old people. You're gonna try and pick out people your own age, ain't you? And you sit down and start talking, you introduce yourself, they introduce themselves. You become friends and you talk. Then, "How are you?" "Why don't you join this?" "We got a beano or whist party." Or something like that. And people working together become friends.

My father used to go to church, and he'd go to Mass, maybe at eleven o'clock on the Sunday. But he wouldn't get home until three o'clock in the afternoon, because when the Mass was over, he'd get out on the porch and talk to all his friends and everybody else. And they'd talk and talk and talk and talk and talk. He'd probably talk for an hour or two in front of the church, with his friends.

G. Do people do that now?

R. No. Now I'll tell you who does it, to a certain extent. You've got a new bunch of people that moved in the city, and they come from Canada, New Brunswick. And of course, when they come out of church they meet. They've lived in different streets. They meet and—all talking that Arcadian French. Then they have a ball.

G. But the other French people don't?

R. No. But the Arcadians are still living the old way, because up in Canada they still have villages. And the village, everybody meets in the village and it's still the old way of living.

G. So that's a very important building for you then?

R. Well, it's important to the people that live that life, anybody's church is.

G. Yes. But to you?

R. To me? Yes. My faith. Yes.

REFLECTING

Raymond's understanding of his contemporary experience was vividly colored by his construction of the past. He recalled his childhood on the far side of the city:

In them days there was no houses there, we were the first house on the street. Then there was a stone crusher. You know what a stone crusher is? There was like a quarry, and they blast stones in the quarry. And then there was a house up by the top of the hill. And then on the other side, from the beginning of the street there were houses up as far as us—small houses. And then there was a man, Ellis, he had a horse and teams, and he done all the trucking for the loom works, and for Stern's and for Vorne's (department stores) and for the stores in them days.

When I walked to school I walked five miles to school. We used to leave the house sometimes at half past four in the morning, jump on the milk cart and help the guy peddle milk: and then go serve Mass at half past seven at Winslow Square: and then go to school. And then at night time, coming home, we used to pick up two hundred and fifty French papers and peddle them on the way home. We never died from it. We were never sick from it, we were always healthy. And you know what the route was? Kids today don't go across the street to peddle a paper. But my route, paper route, would start Harbor Street, Winslow Street, and Pear Street, and Jefferson Street, and come down into Jefferson Square, and right down Waverly Street, down as far as the crazy house. Go back up Sutton Street, up as far as Mill Pond. We'd do that after school; I had two hundred and fifty papers.

The houses, the church, the school, even the stores were remembered:

Come the holidays, we're gonna say Thanksgiving, they had turkeys strung up there, all extra clips and everything else, strung up. And you went in with your folks. That was a big deal. You looked, spun the turkey around, you felt and looked at the turkey. "Oh, I think we want this one." You'd pick that turkey up. "Deliver that." Then you'd order everything that was out, that you think you want. They'd deliver.

It was a happy time and the world was good:

R. What the hell, we played ball, in front of the yard. And if you was on second base, and I was on first, and the other guy's on third, and they knock on the window and say, "It's time to come in and eat," you'd just drop your ball, your glove, your bat, everything, wherever you was—left it there. Maybe you didn't finish that game till two days afterwards. Nobody would steal anything.

When we were children we never broke windows or broke anything.

G. Never?!

R. We got in mischief. My father caught my brother throwing a snowball through the window. He'd just thrown the snowball, and it went "crack" when he comes round the corner. He's coming home from work. And my father, all my father did was say, "That was nice. That was nice. You go up now and tell them that you'll be back after supper to fix that window." See. So my father took my brother and he gave him some money and he made him go down to the farthest hardware store there was—there were three hardware stores between this one and the last one. And he gave him the size of the glass and everything he wanted. No supper. Before supper we had to go down and buy the glass. Then afterwards, after supper—we had supper—he gave him the hammer and he gave him the putty knife and the putty. He made him take the whole damn window off, put it back in, then apologize to the man. Now they don't do that today. They go, "My kid didn't do it."

You know what we used to do? (laughs) We used to have these rag men. You know the rag man? He'd come around with a horse and team. What we used to do. You'd call the rag man in, see, and tell him, somewheres you had a bag of stuff. While you was talking to him, we'd change the bridle on the horse. So when he'd get up, he'd pull the right and the horse would go left. He'd pull left and the horse would go right! And then when one guy was talking to him, the other guy would be taking stuff from the wagon!

G. You were a bunch of crooks!

R. No we'd give it back to him. We'd sell it back to him. (laughter)

Raymond moved to the Winchester Street area in the early twenties. There followed hard times.

R. In the twenties, up to about thirty-five, that was depression time, and in them days what was the most important thing was to try to get a job, to work, because everybody was out of work. That's what we called a man-made depression. Why I say

91

a man-made depression, there was a clique of millionaires that cornered the market. And everybody coming back from the war exhausted their savings and had nothing to live on. So you had. . .some veterans were on the street corners selling apples. Any way to make a buck. I had more damn jobs in them days. After I got out of school, I had more damn jobs than you could shake a stick to. And everybody'd say, "Can't your husband hold a job?" So you'd go and get a job, and they'd say, "Yeah, we got a rush on. I don't know how long it's gonna last." So you'd go and work three or four weeks. Then the order would run out, so they'd have to let you go. They had no more work, just that one order. So they'd say, "Well, we'll call you back as soon as it picks up." Well, you ain't gonna sit on your fat ass and wait for a job, no matter what it is. In them days I worked roofs, sidewalling houses. I worked selling signs, salesman selling signs, advertising signs. I worked driving a truck once in a while for a guy that had a couple of trucks on the road. They needed a spare guy. I worked in a shop every now and again. They had a rush order, they called me—I did anything to keep living.

G. Did you ever think of the future then?

R. No. You didn't have time. All you was thinking of is surviving. Survival.

Eventually, Raymond secured a job with a local bakery.

I built the routes for Allerton when they started. I was like a route builder. For two years you'd go from one house to the other. I had every house from Blyton Lane, that's Taunton Street, and I'd go down Blyton Lane, back of the cemetery, and then down Durham Street, as far as Andrews, as far as the bridge I'd go. Then come down Cutler Street, and then back down Sentry Street. Easthill to Imperial, up Imperial to Andrews, down to Durham. I'd come back down Andrews and go down Easthill Street. There was a couple of houses down the other end. I had a couple of houses on Grant Street. I couldn't go Winchester Street, I had too many customers (Figure IV.2).

So I got to know everybody, every damn one in this neighborhood, every house, every family. See, they all called me, and they still do, a lot of them, "Ray the Baker." I used to take pastry and cut it up in small pieces, in them days people were hard hit, see. And the kids would be playing on the street, and I'd cut them all

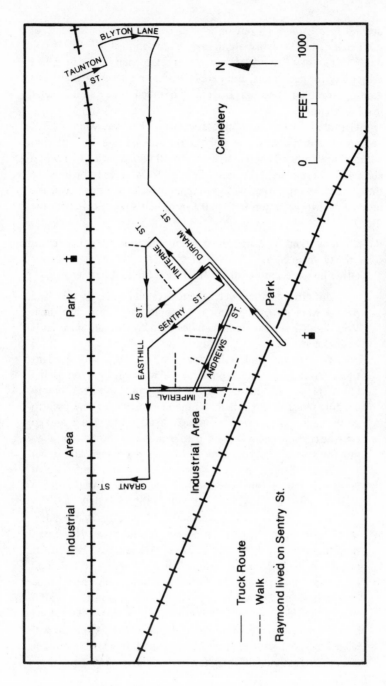

Figure IV.2 Raymond's Bakery Route

93

up and spread them all out on a piece of paper on the tailgate of the truck. Then I'd whistle and say, "Come and get it!" Oh, if you don't think they'd come. I was the greatest man on earth. I had everybody for customers, and people wanted me to go farther but I didn't have time, because I spent too much time talking, see—making friends.

I wanted to be an outside salesman. That was what I liked because I was meeting the public. But then I had to quit that because my father become sick, my mother was sick. So I was a good son, there's me, take care of them. So in order to be close to them, because my father was getting an attack, one right after the other, I had to get a job in the city. And that's when I went in the shop.

Raymond worked as a machinist from this time until his retirement some thirty years later.

Often he would muse on neighborhood life as it had been:

The difference between them days and today is this. The people were struggling, and the people had family life then. You didn't have cars, and families used to visit and play cards with one another.

We'd go out and walk. Sure, you'd take a walk at night time. You, your wife, you'd take a walk and go round the block. Now, you come at night time you don't see nobody walking the streets. Oh hell, in nineteen forty everyone used to go walking night time. You used to hang out the fence and go over next door and you talked to nearly everybody. You'd go out for a walk and you'd go down Imperial Street, and down Winchester Street, and up Durham Street and come down Sentry Street. Before going to bed you'd take a walk, eleven o'clock at night. You'd stop and talk to this one, that one.

Now I could go out on the street, not know the kids, could have ten or fifteen kids playing on the street. Never met them before. "I know you. You're so and so's boy. Hey! Your father's name is Mike, or George." Just the way he acted. You know the old saying, like father like son. You know some people have a habit of saying, "Damn this, damn that, damn this, damn that." And that brings right on. The kid, everything he's gonna say is, "Damn, damn, damn." Or another kid will have a certain walk on him because his father or his mother does that: that's sufficient, they're that close.

94

The aura of the neighborhood past would be embellished and enlivened by memories of notable incidents. There was the 1938 hurricane, a tremendous storm. Raymond was working in the Newsroom at the end of the street:

> The people were getting off the bus, up Winchester Street, and they come through the school yard. And they had their umbrellas. Just as they got by the building, I see one woman being held up, lifting her right up and brought her over and across and knocked her against the fence. So I closed the store and I come home. The paper called me up to go downtown, to the headquarters, because it was news. And I said, "Go to hell!" They wanted me to go down, and they were going to put extras out. But everything was flying. The trees started to fall. We was watching the trees fall, "Bang! Bang! Bang! Bang!" and the wind was blowing fantastic, about a hundred miles an hour. (laughter) Oh, it'd lift you right up. So I said, "Go to hell, I ain't going out there." And a good thing I didn't.
>
> In this house here, right in this house here, we had four trees, on that side of the house, and one tree in the front. They were taller than the house. And that hurricane came, knocked all those trees down. We were very fortunate. Out of all the trees there was only one that leaned on the house.

Invariably Raymond's reminiscences were laced with humor. He reveled in vicarious participation in local pranks.

R. Ooh, we used to light bonfires. We used to take old tires and put them over the hydrants, and set it afire. And the fire team would come, and they couldn't get the water! I'll tell you the funniest one we had. There was an old guy by the name Baker who was a baker. He had a baker's shop on Durham Street between Cutler and Sentry, and he had some steps. Now we were building a bonfire—*not me, the kids*—down the corner of Tinterne and Cutler Street. They had old boxes. And he come out of his house with a hose. The kids jumped the fence, and they let him get near the fire, start putting the fire out, and they cut the hose. They went down and stole the steps. They said, "Here's your stairs," and they threw it right on top.

G. I thought this was a nice neighborhood?!

R. It was, those days, but that was the Fourth of July. That was a lot of fun. Oh, we had fun in them days. That's all the harm we did. We never broke anything.

He paused. "This is the old society."

INVASION

Memories of the "old society" contrasted with Raymond's view of the neighborhood of the present. He was sorely troubled by the physical decay and social disintegration of the neighborhood. Yet he seemed somehow sanguine, sometimes almost a detached observer, on occasion eager to vehemently rue the changes, but at the same time able to disassociate himself from what was going on around him. One day we were talking about the boarding up of stores on Imperial Street. Abruptly, he jumped up from the table. "I'll show you," he said. I followed him onto the rear porch. It overlooked a building supply storage depot which had been encroaching into the neighborhood. A fork lift truck was busily shunting piles of lumber. From this vantage point we were afforded a view over much of the neighborhood. Standing there in the misty grayness of a mild February afternoon, I was given a lesson in urban geography.

Down the other end, that's all new stuff down there (he pointed in the direction of the city center), that's all new stuff. See, that's, that's changed so very much there. The people have let go. It's changed hands and they've got the Puerto Ricans and stuff coming in down there, and they're destroying it. This used to be beautiful at one time (he gestured towards Imperial Street). This used to be—it was quite a place. Winchester's all shot, too. You go down on Andrews Street and up the other end of Imperial Street and all that stuff—Imperial Court, Serin Court—it's all decayed. Now that's the blight, what they call the blight of the city.

Well, now in the last, we'll say in the sixties, the Puerto Ricans come in; and it's the undesirable Puerto Ricans. What I mean to say, when anybody migrates—in our days the Italians down through Waverly Street—most of those migrators were from Sicily, or a couple of other countries, and they were wanted for murder in Italy, and they were fleeing over here. So they had connections and come in. The Puerto Ricans are the same thing. They're dope peddlers and they robbed.

He explained that the Puerto Ricans were responsible for many recent fires within the neighborhood:

Well, they smoke in bed, and they're under the influence of dope. They're in trouble all the time, so they cause an undesirable neighborhood. The people move away—the people, the original people that have been living there because their folks lived there. The undesirables have moved in as far as Imperial Street now. You see, the Jewish people buy out the houses when the family gets down, and they rent it to anybody they can get rent off. They rent it to these people. They put in Puerto Ricans. I'm not against Puerto Ricans. There's good people and bad people whether they's white or black or Puerto Rican, but we happen to have all the bad types.

Now your city government is rotten, to a certain extent, because they don't enforce the laws, and they let them expand and expand and expand. This lumber company [we turned our attention to the activity below], it was a little bit of an office, used to be like a brick yard. They used to make bricks there. When he bought out, well, he was going to put a little bit of lumber there, going to sell lumber. Well, they had some houses there. So he bought this old lady's house—and how did he get to buy the old lady's house? Well, his lumber used to come in from Maine, and he'd take a fork lift truck and put it all on the sidewalk and block their passage. If you was coming out of your front door and you had to climb over a pile of lumber, what would you do? You'd get discouraged, "Ah, the hell with it! I'll get rid of it." And you'd sell it to the Jew. They used to have a nice little house over here with shrubbery and everything else. They took the timber: piled it right on top of the shrubbery one day. Lady called the police. They make them move the lumber but they never arrested them or bring them in court or anything else. No sooner had the police gone, they put it back there. Well, they sold out, and it's been that way, and they've been trying to do the same thing with us.

He pointed down the street.

R. The house on Garrison Street here, on that side, at the corner, he [the original owner] was a mason and a plasterer. And when he died, his widow. . .after a few years, the Jew offered to buy it and let her live there as long as she lived. Now she sold to him. But what happens? She was an old woman, so he says to her: "You move upstairs, we'll make an office downstairs." Then, the first thing you know, he forced her in her room.

G. Into her room?

R. Yeah, one section of the house; just the kitchen and the bathroom.

G. When was this?

R. Oh, this was back fifteen or twenty years. So what happened? What happens? The lady dies of grief. She sees the mess they make.

There was a long reflective pause, a shared feeling of quiet sadness. We stared out over the ruins of the once vibrant neighborhood.

G. How long will it take for the decay process to hit this street?

R. Who knows, who knows. The pressure's on. Five years? Ten years? Fifteen years? This neighborhood in another twenty-five, fifty years, won't exist. It'll all be rebuilt. As the neighborhood is now, it's marked, it's marked on the plans, down city hall there. It's marked for renovation. When? How long it's gonna take? I don't know.

They take over and the city lets it go that way. You can't borrow money from the banks to buy houses down here. They don't do nothing. They don't do nothing to help.

I asked if there was anything the community could do.

The French people are quiet people, they're a peaceful type of people. They don't want to be involved. In order to do something you've got to get involved. People get tired of fighting. You'd have to have a leader. If I was younger, I'd do it.

He began to talk about personal experiences with the encroaching lumber yard below us, as we stood watching the fork lift truck:

I got mad. I went down there. I said, "Memorial Day is coming, Friday," I went on the Monday, "and I want you to clean that street up." "All right, all right, we'll clean it." So Thursday came: the street wasn't clean. I comes in, drives the car in the driveway, goes back down there. I told him, "You're supposed to clean the thing, and this is Thursday and tomorrow is Memorial Day." He says, "I ain't working for the city." He says, "We ain't working for anyone, we're in business to make a profit." I says, "Son, if that's not clean, I'm getting a crew together; we're going to clean up, and going up and dump it on your front piazza." No sooner get into the driveway and he says, "Hi! Hi!" And they got it done.

I had the inspectors down a dozen times, but you can't get anywhere because there's too much politics in it. They don't bother me too much now because I've been after them fifteen or twenty years. Well, at least I've done my part. It don't mean anything to me now. My time, my days are numbered, you know what I mean. The few years I'm gonna live here.

It was becoming colder and the wind was picking up. The clothesline on the floor below rattled ominously. We moved inside. I asked Raymond about the changing social fabric of the neighborhood:

R. When you're gonna get people like this living all around you, stealing everything you've got, breaking everything you've got, nothing's done about it, well, your kids say, "I don't want to live here any more." They move out. They ain't going to live around the neighborhood. "We're going to go over to Orion, Wexford, Cornwell, Waverly." [suburban communities] The city at one time had two hundred and seven thousand people. Now it's down to what? A hundred and sixty-five, hundred and sixty-six thousand. Where's all the people gone? They've been pushed out. It's pretty hard for any man to have his ties when he knows he's gonna be pushed out.

G. But a lot of older people are staying.

R. Well, it's because, what the hell, what are they gonna do? Where are they gonna move? They stay because their friends are here, the old people are here, they can communicate. Now if them old people all moved somewheres, and they weren't here, then they'd be lost, because you don't talk the same language as your kids.

G. Do you think that you'll have to move from here?

R. Well, the plans are down the road. Like I say, what the hell. When it becomes no more safe to be on the street, to walk the streets or anything else, then you don't dare go out. I don't go down and get the paper or anything else, not unless I go in the car. If I needed a loaf of bread I wouldn't go out. You know that kid that was picked up for murder?

G. Yes.

R. Where does he hang out? Down the corner here.

G. Do you ever think of leaving because of it?

R. Well, if I was young, I would. I'd get the hell out of here. I'm just sentimental because it's a family house, you know. If I was younger I'd get the hell out.

G. So you never want to leave this area?

R. Oh, I don't say that. I may. But you kind of reluctant. Now if I didn't have this kind of association with my nephews and nieces? Now this is a family house. I'm the last of the old people left. Well, I may go. I may go off to my daughter's in Arkansas.

G. But you don't want to leave now?

R. Well. What the hell. I'm too old now.

LIVING FULLY

Often our conversations would focus on Raymond's philosophy of life. He was an active vibrant person, yet he acknowledged his limitations, subtly juxtaposing a knowing realism and an unshakable faith. He was at one with himself in a context where others might despair. I was eager to learn the secret of his fortitude and good humor.

R. We're supposed to be just passing through. Death is supposed to be a glorified thing because you're going for your reward. So if you analyze the whole thing, you're not down here on earth for monetary, for pleasurable things. You're given time to make your way, prepare your way for your reward that's coming. So, by that same token, live day for day, do the best you can. You don't know if you're going to be here tomorrow or not. You live the best you can, you have no regrets, you have a clear conscience. You take your bad with your good and your good with your bad, and you make the best of living. Cause you're put on here for a purpose. And if you throw that purpose away, defy that purpose, what good is it? That's why I am always happy. That's why I never begrudge anybody. I never begrudge you. You're my neighbor. All right, I work like a son of a gun. You don't do anything and you get a million dollars. Boy, good luck to you. I don't begrudge you. "Why, that son of a bitch gets a million dollars and I didn't." No. It's meant that way.

I don't know how long I've got left. You don't know. Five years, ten years. You're on borrowed time; well, once a man reaches sixty-five he's on borrowed time. Everything over sixty-five is good.

100

How the hell do I know I'm going to wake up tomorrow morning? How do you know you're going to wake up tomorrow morning? How the hell do you know you're gonna be able to go out tomorrow? How do you know that we won't have six foot of snow, or if it rains so hard it'll be all icy? You go down, fall down, break your leg. So why worry. Wait till tomorrow morning, until tomorrow morning comes. "Well, good Lord, I'm here." All right. Now I'm gonna figure out what I'm gonna do. Ten minutes time you think it out.

G. Have you always felt that, or was there a time when you were not as wise?

R. No. No. I was always a type, never worried about that. My wife was the worry wart. And I used to say, "Put a shingle up," you know. "Put your name up and say, 'I'll worry for you,' buck an hour." I used to tell her that. She used to say, "Ah, you never worry, you never worry." And I used to say to her, "What the hell good is it for me to worry and you worry? You make yourself sick. I'll be healthy, then I can take care of you. But if we both get worried and both make ourselves sick, then who's going to take care of me and who's going to take care of you?" I used to tell her that.

If you was to think of every possible thing that's gonna happen, you'd be in a nut house.

G. But don't you ever make plans for the future?

R. Plans for what?

G. What you are going to do next week, or in the summer, or. . .

R. No. I'll get up and I'll do it.

G. Did you always used to be like that?

R. Yeah, take things day for day; you plan something, and then what happens? Something comes up. You can't do it. All your plans are gone. And I've always found that the best way for a vacation, the best way to do it is just on the spur of the moment.

Me, I do everything in order to have fun. I like to have fun. And if you've got fun, there's no tension and you're less apt to have a heart attack. You see, what brought my heart attack on was too much pushing—work, work, work, work. I've only got a small income; I live within my means but I enjoy myself as much as anybody could. Now if I had a lot of money, I'd probably be dead this year! Because I've got to go here, go there, go here. (uproarious laughter) I'd try to keep up with the Joneses. You've got to remember this. I ain't no spring chicken any more, and I have to stop and think now. I don't push. If I push, it'll catch up with me. You know what I mean? I can do anything except physical. But if, like in the winter time, if it's stormy, cold, I don't go out.

My sister-in-law, now she lives alone. And I said to her one time, "Why the hang don't you join the old folks club at the church, and go on the bus trips with them?" So she went over. They was going to have a trip, and she went on the trip. But what was her reaction when she come back? She calls up. She says, "I'm not going any more." I says, "What?" "They're all a bunch of old biddies." I says, "They are? Well, who are the old biddies?" So she named them. I know them all. I says, "Liz, how old is Mrs. Jones? Seventy-three? How old is Mrs. So-and-so? Seventy-four? How old is Mrs. So-and-so? Seventy-one?" And I said, "Look, how old are you? Seventy? And they're old biddies!? Don't you forget that you're on borrowed time, too." See, she was wrapped up with her daughter. And her daughter's one of those go, go, go, go all the time girls. She's full of piss and vinegar. "Ma, I'm gonna pick you up. We'll go to the horse races." "Ma, we're gonna pick you up and go to the beano." "Ma, we're gonna pick you up and go this and that." Well she wanted to be young, which was no good. It was killing her, in a way. But she wanted to keep up with the young ones. Well, you get to the stage where the young's got to go with the young. The old's got to go with the old because you talk the same language.

Raymond seemed too complacent, too satisfied, so much in control. My incredulous skepticism would provoke me to probe for signs of disenchantment. I asked if, when he was younger, he had anticipated growing old:

R. I always lived day for day.

G. Even then? You never thought: "Well, I'd like a nice little house in the country," or anything like that?

R. No, because I always figured this: when the time comes, if I could get enough money ahead, I'll do it. But you couldn't do it, you couldn't get money ahead.

102

G. What would you do if you got enough money to get ahead?

R. Well, I'd, I'd like to live in a little house, a little place in the country *with a garden.* I could potter around. That's what I'd like to do, I would have liked. But I never did since I always had some obstacles in the way. See, when my folks got sick, then her folks got sick, her mother died, then I took in her father. Then he died at ninety-six. Well my life was pretty well established.

G. If you wanted to live in the country, why are you so happy even though you haven't? You're not living in the country now.

R. You make the best of the situation. We all have plans. You have plans. I have plans. Everybody has plans. But them castles fall down, don't they? And because the castles fall down, what are you gonna do? Are you going to take a gun and shoot yourself?

G. No. No.

R. Now it hasn't improved in the sense that I could have made it improve. I could have had my own home, and I could have had my own this and that, if I'd saved all my money.

G. Why did you decide not to?

R. I put my son through school, my daughter through school. Now if I was like a lot of people, "Go and get your own money if you want to go through school." I helped them because it was family and it was good. That's what families are about. Help one another. It's not the toys that you buy. It's what you can give in education. And my ambition was that they wouldn't have to work as hard as I did. The aspiration was to give my children the opportunity to earn a better life than I had. I had to struggle for life. Now I sit back and I don't have to worry. I don't worry about my children because I know they are both well off. So I know that I've achieved my goal. They earn a better living than I had.

I'm relaxed. I don't give a damn. I never cared for money. Me, personally, I never cared for money, whether I had a dime in my pocket or nothing. I always enjoyed myself. What makes people happy, what makes people satisfied, is if they've got families and the young ones and, you know, immediate families to relate with. That's the basis of all happiness of old

103

people. You raise a family and if your children communicate, and live with you, I mean, and relate with you, what more do you want? Your success after you've raised them *is to watch their family grow, and take part of it.* You don't want to be in their hair again, but the people who haven't got anybody? I adapt to my situation, understand and accept it. Some never accept it. Some never accept that their children leave them, cause they've got to leave.

I'm living on borrowed time. What the hell. You gonna die, I'm gonna die, we're all gonna die. But if you're so worried, you ain't gonna face it. Why worry about it? Just say, "Well, I'm gonna die. I lived my life through to seventy-six." You sowed your wild oats and everything else when you was kids and you had what you wanted. Okay, your life has ended. Your life is gonna end, so adapt yourself to it.

G. Do you ever think about that?

R. Me? I don't think of it. When the good Lord says, "Follow me," I'll go. Whether I have a "boo boo" or whether I don't have a "boo boo," you got to go anyways. Like the saying goes, it's the only thing that's certain. When I had my heart attack I was in intensive care. I had a galdang tube into my heart, and that there thing they had up there with the spigot. They used to tell me, "For Christ sake, why don't you shut up?" I had more damn fun when I was there. They said, "Don't you know that you're going to die? You can die." "So what. What are you worried about? If I die, I die." I couldn't do anything. If I cry, it's gonna make me just as sick. I may as well die happy as die crying.

G. Yes. But haven't you ever worried about that—ever? In the past?

R. No, because I'm only passing through. Everybody is just passing through. You're born. You've got a span of life, that high. Okay. It gets narrower and narrower and narrower and narrower. When the top and the bottom meets, you're gone. My life is numbered.

G. Numbered?

R. The days are numbered. So you may as well enjoy. I don't know what number I'm gonna get to. I want to have a good time while the numbers are coming! What the hell. It's just like that saying. Doctor says, "You can have just one more. I want you to cut out smoking." "I can't." "Well, look, you can have just one pipe full of tobacco." "Okay, just one pipe, that's all I ask." So his doctor goes away. He goes down the cellar and gets himself a big wooden barrel; makes a hole in it; puts a galdang pipe in there, and makes a stem out of it; fills it full of tobacco. Doctor comes back—a couple of weeks. He's still smoking that thing. "I told you to give up smoking." "You said one more smoke."

CHAPTER V

EVELYN

"I'm happy that I've got my children and happy that I'm still alive."

There is a small name plate on the door of the cottage—Andre Renault. Andre no longer resides here. He died sixteen years ago. Evelyn, now seventy-six, lives alone.

He always lived in East Lanchester, and I met him in East Lanchester. I can remember when I was single, my husband living in this house. I used to see him, and I used to think he was nice, you know what I mean, when you have your eye on somebody. And then later on, we went to dances, and that's how I met him. Course, years ago we used to go out longer than what they do now. I went out two years, two years and a half with him before, because work wasn't so great at Restic and Fulton, you know. He was working there. So we delayed getting married.

Well, he went to work on a Friday afternoon, three to eleven. And he came home and he says, "I'll have toast and some peaches," and he went to bed. About two o'clock in the morning he was all sweat, and pains across the chest. That's how quick he went, never realizing it was, it was really a—I had no idea it was a heart condition.

Why did it have to happen to me? I just thought I was being punished. But it happens to others. I didn't know what I'd do alone; but it's just something that everybody has to live with. I think you realize then. One goes. Maybe you could go, not very far after. But I've lived sixteen years.

Even now, some of her actions reflect subservience to his philosophy. She has no checking account. "My husband didn't believe in them. Too much work, he'd say." However, today Evelyn is very much her own person. Well-groomed gray hair, her slightly rotund figure, her unhurried movements, and her kindly grandmotherly manner, convey an impression of well-being. Almost effusively friendly, yet quietly assured, she is open, trusting, inquisitive, eager to help. A tolerant equanimity seems to stem from an acceptance of the limitations imposed by advancing years and calm acknowledgment of the uncertainty of destiny. Unlike Marie, she never rages.

"I'M CAREFUL."

Evelyn's modest Social Security income is supplemented by a small railroad widow's pension. "I'm medium class, we'll say middle class. To tell you frankly, I'm satisfied." Nonetheless, she is obliged to be frugal. Taxes on her home, the rising cost of utilities, and the burden of health insurance, consume much of her income. Once regular expenses have been met, the remaining discretionary income is more flexibly assigned.

> I go along as I need it. You have to know how to manage. If one week I feel that I can't spend too much, well, being alone, I know enough to buy—I don't need eggs every week. I don't need butter every week. You can stretch, you know. And I watch the sales a lot. If there's any coupons, I watch it.

The furniture is old, the carpets are slightly worn. Rationalized maintenance takes precedence over acquisition.

> I'm careful, because I think, what's the use of always getting new things when you can get along with what you've got that's livable. That's the way I think.

"AND I ENJOY WHAT I DO"

She lives quietly. A normal day is a mixture of household chores, television, a visit from a relative, and perhaps a local excursion. Arrival of the evening newspaper, a telephone call, the unexpected caller, are key events. Yet Evelyn remains judiciously active.

> I always find something to do. I don't try to sit down and have nothing to do. They tell you, many of them tell you, the more you keep going, the better it is for you. I think that's true.

So she is constantly dusting, vacuuming, sweeping the step, polishing, and "picking up" around the house.

Long periods during the day, and most evenings are spent viewing a large color television.

> I relax at night listening to TV, right there. I can have a part of my night there. I think it's like medication, or what other word I could use? Therapy. And then when I'm tired of there, I just go to bed. I sleep downstairs. I leave the upstairs.

108

In recent years Evelyn has become increasingly dependent on relatives for transportation. "Well, I go out less." There was a time when she would regularly walk to the stores on Imperial Street, stroll through the park with her children, or wander even farther afield. "I used to walk down city. Now, I don't think I'd attempt it."

A combination of factors explain this withdrawal. The demise of local stores has reduced the number and attractiveness of services within walking distance.

E. We used to shop on Imperial Street when things were better, ten, fifteen years ago, but now every time you go the price is different.

G. So you never go down there now?

E. Unless I'm caught without, you know. You have to if you're caught.

Fear of victimization has discouraged walking.

E. We used to enjoy taking a walk up the park, too, when the kids was small.

G. You don't do that any more?

E. No. I'd be afraid now. You know it's dangerous to go any place, even in the daytime.

G. So you don't go out in the daytime, alone, ever?

E. Well, no, not a great deal, unless I go just to the store.

G. Do you feel scared if you go to the store?

E. Oh, no, no. I don't. I try not to think of it that way.

In addition, she acknowledges the constraining influence of environmental conditions. "I wouldn't mind walking if the weather and the sidewalks were good." Finally, self assessment of reduced capability, "It's a little too far to walk," in conjunction with external sanctioning of non-participation, provide justification for her reluctance to venture forth. With regard to church attendance, she explained:

Like this winter, when it was slippery and if Lucy [her sister] wouldn't have been around I wouldn't have gone. Oh, no. You're exempted from that. They don't ask you to do the impossible. You have your services right on the TV. No, at my age, I want to stay in. I'm comfortable at home. I'm comfortable.

109

It would be erroneous to conclude that Evelyn's life-space has become uniformly constricted. On occasion, the excitement of a family wedding, a graduation, a new birth, or the tragedy of a close friend's passing, punctuates the rhythm of daily life and provides the opportunity to leave her home. Limitation of her daily activity orbit is also partially compensated by occasional long distance trips.

When her husband was alive, Evelyn made three trips to Florida.

Well, he was a railroad man, and with him being on the railroad, I could go with him, once a year—and he'd take his vacations in February, two or three weeks.

Andre's death did not lead to curtailment of such excursions. She explained how the friends she had made in Florida had been an important source of support when she started to make the trip alone:

They made it nice for me. I went down twice by myself and meet them, they met me.

In fact she seems to have acquired a taste for travel:

Oh, I've traveled more since my husband's gone, too. That makes a difference. Well, I think I went three trips with him. He showed me the way around. But, I mean, now I go with groups because it's easier.

Sometimes she describes a car trip to Canada. She has also been on an excursion to Washington. "I went with Mrs. Cormier, my good friend. We went two busloads."

The most recent venture has been a journey to visit her son in Arizona. "My first time I ever flown, too!" During her stay she was taken on several car rides.

I went to Mexico, even. I didn't go right in the center of Mexico—we drove about two hundred and something miles.

Her next visit is eagerly anticipated:

Well, I'd like to go if my health keeps up. I wouldn't mind going back again, but I can't go back right away. I'll wait till next year, I mean this year, later in the fall.

Vicarious participation in a larger world also compensates for reduced physical involvement. The television, so important in Evelyn's life, conveys her beyond the neighborhood. East Asia, Washington, California, all are accessible through the media. The personalities on the quiz shows, and the newscasters have become an important part of her world. She can share in the tragedy of a Dallas afternoon:

110

I was here in the afternoon. I can remember sitting there, and the news came on. And I have a friend that called. She's deceased now. And she says, "Did you hear the news?" She said, "Our President was just killed." And I put on the news and, you know, this one that's on Channel 9 [Walter Cronkite], he stood there, "How could that have happened?" I can remember him. He was almost sobbing. He was.

She can also monitor the less sensational:

I hear on the news last night, I don't know if it's a bus accident, or whether they were on the bus, but two children from Connecticut are in critical condition. I didn't quite get it. I'll follow the paper tonight. Maybe I'll get it better.

The evening newspaper, a complementary medium, facilitates participation in the local scene, and avid reading of obituaries maintains a sense of affinity with the ongoing pulse of the community:

You get your friends, and you think, "Oh well, who's gone?" Isn't that funny? The obituaries, it's the first thing you look at. That's the first thing I look at.

"THEY'RE VERY THOUGHTFUL"

In part, Evelyn's equanimity stems from her perceived role as matriarch, the fulcrum of an extensive family network. This set of linkages provides her with a sense of significance within a supportive group upon which she can depend. Figure V.1 represents a portion of her family tree. As we traced the linkages, I was constantly amazed by the unerring accuracy of her recall of myriad birthdays, ages, wedding days, and dates of deaths. Many of her closest relatives still reside within ten miles of her home. Her sister, Lucy, and two sons are frequent visitors.

He was here yesterday, my oldest son. I tell you, I have company pretty near every day. Sunday I had Lucy, she stops in every other day. Now yesterday she didn't stop, but she might stop today.

A close relationship with her third son, in Arizona, is also maintained. She often talks about his home and family:

Oh, it's lovely. He's got a nice little place. Yes, a single house, but it's a nice little place, and everything is so level and they're so adjusted. Oh, the kids are so adjusted to everything.

111

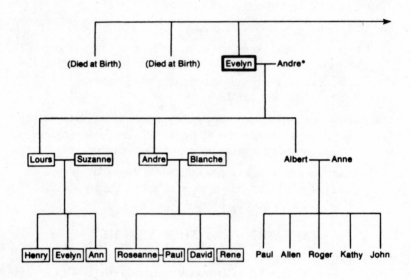

- Deceased

☐ Resides Within 10 Miles of Evelyn's Home

Figure V.1 Evelyn's Family Tree

112

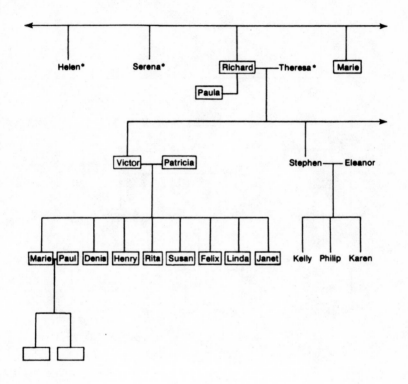

Figure V.1 Evelyn's Family Tree *(continued)*

113

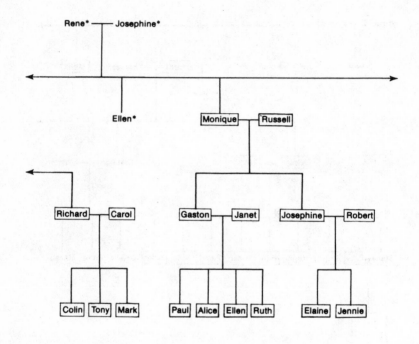

Figure V.1 Evelyn's Family Tree *(continued)*

114

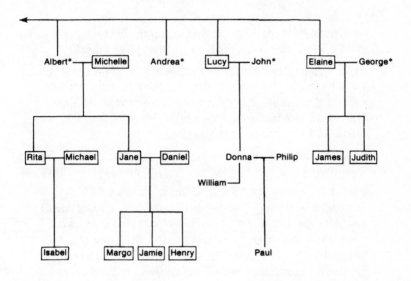

Figure V.1 Evelyn's Family Tree *(continued)*

Exchange of letters and photographs, physical reminders, including a tray with a map of Arizona emblazoned upon it, and several books with photographs of the countryside around her son's home, supplement more immediate vicarious association:

> I watch the map on the TV, and they've got cool; and it's down to about, like in the forties, forty-five. He's in the center of Arizona.

Evelyn strives to maintain her role within the family but, characteristically, her efforts are tempered by a cautious pragmatism:

> I was supposed to have them here for a New Year's party and get-together, as the mother. And I couldn't, see, I get tired easily. I don't mind doing a little work for myself, or three or four of us, but that would have been fourteen, with my two sisters. So I took them out. I splurged.

Love for and pride in the family is frequently manifest. She recently received an expensive ring from her children, as a birthday gift.

> I don't know why they did it, Graham. At my age. I said, "I don't need anything." I really don't need things. I buy what I want. You're not a kid at over seventy, you know what I mean, those are things I can give up. Oh, that's the way they feel, so let them be. I said, "Well then, when I die, don't forget to take all them things away." I don't want to be buried with all diamonds and rings. I want them to leave it to the children.
>
> Sometimes, you know, when you get old they don't care whether they come or not. I mean there are some like that. But I can't say that. Mine are very close to me. I don't care if they only, even only took me out for a meal. I'd be satisfied. But they're very thoughtful.

Quietly, she revels in the family's attentiveness.

Paradoxically, the potential availability of family provides justification for her resolute independence. There is a trace of stubbornness, a concern that they might be overprotective. On one occasion when I called, I found she had taken a nasty fall. Her leg and arm were badly bruised.

E. Oh, that's all right. You get many bumps in a life.

G. Doesn't it worry you, being alone, when something like that could happen?

116

E. Yes, I know. And then you get over it and start all over again. (laughter) I try not to tell them, not to scare anybody. Lucy knows it, and then my sister knows it, the one up on Indiana Street.[1]

G. What did she say?

E. Oh, she said, be careful; but it's just one of those things.

"I STILL LIKE TO BE NEAR MY CHURCH"

Evelyn's faith is an unquestioned and unquestionable rationale rather than a deeply reasoned spiritual philosophy.

> It's like believing in one God, because if you don't believe in nothing, what are you going to believe in? Don't you think? I think so, anyway.

She is strongly committed to her church:

> I don't want to see it destroyed. Attached to it, that's right, like you would be attached to your father and mother. And that's a place to go until we're not able to move, or sick, or something like that. I give a dollar a week. I think we owe it to our church. If you went to the theater, then you'd pay your way in. You're going there for the good of your mind and, I don't know, for the help of your pastor.

The church, symbolic focus of the French community, and a meeting place for clubs and societies, constitutes the hub of an extensive friendship network. Intimate knowledge of local events emanates from this more secular attachment.

"YOU GET YOUR NEIGHBORS AND YOU FEEL AS THOUGH, AS THOUGH THERE'S SOMEBODY CLOSE TO YOU."

Many of her friends are age peers (Figure V.2), but a more significant distinguishing characteristic is their membership in an essentially

---

[1] It is noteworthy here that Evelyn only informed her peers, not her children.

117

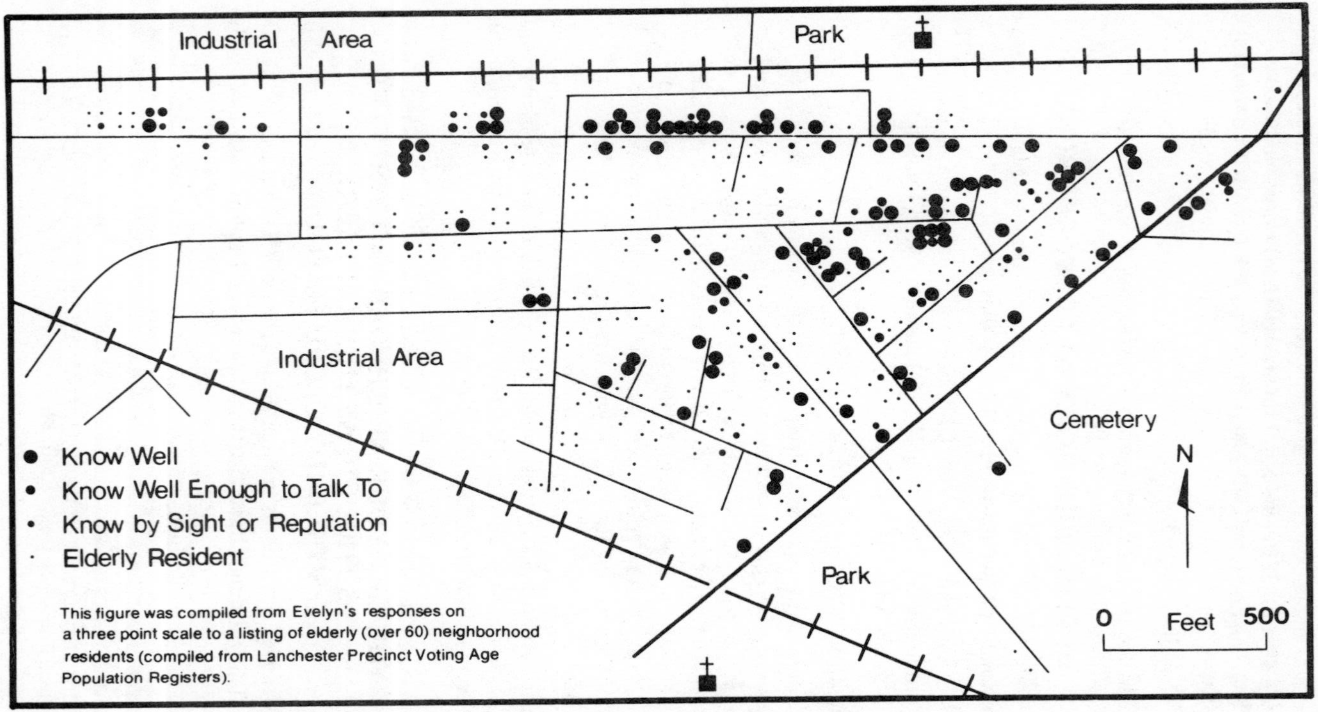
Industrial
Area
Park
Industrial Area
Cemetery
Park
N
0
Feet
500
Know Well
Know Well Enough to Talk To
Know by Sight or Reputation
Elderly Resident
This figure was compiled from Evelyn's responses on a three point scale to a listing of elderly (over 60) neighborhood residents (compiled from Lanchester Precinct Voting Age Population Registers).

Figure V.2 Evelyn's Winchester Street Peer Group Social Network

French Canadian cadre.[2] It is thus a selective friendship network. "See, there's a lot of them (the non French Canadian) I don't know."

Her supportive social network provides Evelyn with an extended neighborhood "family." There are many endogamous linkages.

> I have my brother-in-law's brother married to Mrs. Duvalle's (Marie's) daughter. See, Blanche, my daughter-in-law, she was a Coutoure. And her brother, Coutoure, married Mrs. Duvalle's daughter, Suzanne Duvalle. It don't make me related with her but it's a connection.

There is also a quality of surrogate familism stemming from a legacy of lengthy spatial proximity and a sense of community identification. She is involved in the ongoing social history of the neighborhood, grounding her sense of belonging in a rich mosaic of remembered social structure.

> When I lived at 133 Winchester Street they were our neighbors at 135. See, the McNamara's were our age and we were all single. But they got married and push down here, see. And O'Brien's store, well, she's a McNamara girl. And these are her brothers that are running this now. But the package store, it's still in her name.

Evelyn "participates" in the development of local relationships—a birth, a death, a forthcoming marriage:

> I worked with her mother, with Jeannita's mother, so there was a long acquaintance there. It surprised me when she said she was going and—get married—to him!

An almost maternal watchful caring extends even to the paper boy. "His father used to go to school with my kids."

Evelyn diligently maintains her immersion within the social fabric of the French community. As peer group friends die, relationships with younger substitutes, often neighbors, intensify. However, less time is spent in direct contact with those beyond the immediate vicinity of her home.

> I don't go out to see many people around here. No, they come in to see me. Oh, they don't come too often. Or I see them at church, or something like that.

[2]A strong identification with her French cultural heritage, focused on the church, was established early in Evelyn's life:
And then I went to Sacred Heart Church down in the center there. I went four years there. I went four years when they had the school on Pear Street. Well, my father wanted me to learn French. In the public school you couldn't learn French.

119

Indirect forms of linkage through an elaborate informal information network have assumed increasing importance.

> Well, you read about the ones that have gone; and we hear about them being sick, and that's how we find out. Oh, I'll be outside here, like when Mr. Belisle was sick. He died of cancer. He went in a couple of times, in the hospital. Conversation with the neighbors, or Mrs. Royale (opposite) would tell me sometimes of things.

The telephone is also an important medium for the exchange of news and gossip.

> Mrs. Duvalle is older than me but I don't visit her. I have talked to her on the phone. I have a friendly talk with her. And she'll call me if we have a party or a supper or something like that, and, sure as you're living, when she calls we have a good chat together.

"I THINK I'D BE LONESOME
IF I MOVED AWAY FROM HERE"

In many respects Evelyn's sense of involvement within the local setting is expressed in the way she subconsciously differentiates her environmental context.

Home is a special place. It's very familiarity, stemming from forty-one years of residence, furnishes an aura of comfort and belonging. It is space where Evelyn is in control, possessed space in which potentials and limits are known. The bedroom window affords good lighting for reading, the edge of the carpet is scuffed in the living room, and care is needed to avoid tripping. Each nook and cranny has its own character as negotiable space. However, it is possible to unrealistically romanticize the meaning of home for Evelyn. Her home is not a museum. Certainly there is a favorite chair, a treasured cabinet, a set of steak knives with special meaning, even an array of artifacts which trigger reflection on a vivid collage of remembered experiences; but Evelyn's identification with home is, in essence, more pragmatic than wistfully sentimental.

G. If you were to move tomorrow and somebody said you could take two things from the house, what would you take?

E. I'd take a bed anyway! (Laughter)

G. Assuming that you could have everything that you need, what things would you take because you're attached to them?

E.   There's so many things that we're attached to. I don't know.

G.   What things are you attached to?

E.   I like my sewing machine. I love my sewing machine. I don't think I could go without a sewing machine. No. I make up curtains and things like that.

G.   Is there anything that has sentimental value?

E.   Well, I've got steak knives that my husband gave me, half a dozen. Those are nice. They claim those are English. You want me to show you?

It took her almost twenty minutes to find them.

More importantly, home is womb, a place of retreat, closed-in space, secure from the ravages of the alien outside. It is in the sanctity of the "in here" that Evelyn's affinity for and demarcation of home space resides. "I'm never too afraid *as long as I'm closed in* and the doors are locked." Territorial integrity must be maintained.

> I don't know whether they know I'm alone. You hate to ad-
> vertise that you do live alone. I don't think that you could do
> me any trouble, but I mean—if somebody comes at my door, I
> don't like to say. I say we live two; and it's a white lie because
> you don't know if they're gonna try and come in later on to
> see you or something.

Outside, the field of vision from her home defines an aureole of partial possession and shared responsibility. The inner boundary of this zone is the threshold. The outer limit represents more, however, than the farthest horizon of visual acuity. In addition, it marks an important qualitative transition in the meaning of space: in part construed as the limit of encroaching physical deterioration and social invasion, and partially experienced centrifugally as the farthest spatial extent of a sustaining social intimacy. Within this zone, "We are trying to conserve what we've got."

In the summer Evelyn will sit on her front porch benignly monitoring the life pulse of the immediate vicinity. Sometimes we sat together musing on the events transpiring on the street below. Evelyn seemed to know most of the children and many of the adults who passed by. We talked about the inhabitants of many homes within our view. She knew surprisingly intimate details of their lives. Her kitchen window overlooks the school yard.

Over the sink there, that's a nice window. I can see the children in the recess, or else if I'm doing my dishes, people getting on the bus, or something like that, I can see. I like to look out there. And sometimes I wave to the kids, you know what I mean. They're cute; they're not all bold, you know.

Within this surveillance zone, physical change is sharply perceived by Evelyn and viewed as an indicator of significant social transition. We were standing gazing into the school yard:

This school here used to have a bed of flowers—now I would say as far as that tree there. Yes. There was a fence there. They had all kinds of flowers. So that would protect my house. But now that they have no flowers, no nothing, well, they're that much closer to my property.

She pointed to broken windows on the third floor. The school department had placed a grill over the first floor windows more than a year previously. The second floor had been similarly protected this spring. Evelyn expects the third floor windows will be shielded soon in response to the vandals' success with progressively higher targets.

The space around Evelyn's home encompasses a mutually beneficial set of neighboring relationships characterized by both poignant reciprocity and inquisitive although cautious surveillance:

I have Mrs. Royale. Her sister came over yesterday. She was sick, she was sick, Mrs. Royale. So I sent her some chicken soup. I made chicken soup and sent it over with my son.

I told her [Mrs. Royale's sister] how you come. She noticed you, you know. She wanted to know when you were coming. They ask me, well, who's my friend. And I said, "Well, he's coming over today." You don't want people to think that—you know. And I said, "Well, he's, he's educating himself, you know." "Oh," she says, "Yeah, well." "Why don't you come over?" She didn't want to interfere.

The surveillance zone is embedded within the larger realm of Evelyn's neighborhood. Neighborhood space is less intimately possessed but, nonetheless, qualitatively differentiated from space beyond. The boundaries are indistinct and vaguely sensed, a fuzzy coincidence of parish periphery and physical breakpoints (the railroad tracks, the cemetery, a cluster of stores).

There is a seemingly contradictory dualism underlying Evelyn's awareness of neighborhood space. On the one hand it is the parish,

122

focused on the church, a socially coherent and friendly area where people are known—the arena of the French community with which she so strongly identifies. But this space is also chaotic space, pervaded by a motif of degeneration. Here she echoes a familiar theme:

E.   This has changed ever since my husband died. The stores were beautiful around here. We had a nice pharmacy here, Lebouf's drugstore, and that was lovely. But see, well now it's Selena's, they have that nightclub.

G.   I'm wondering what will happen in thirty, forty, fifty years time.

E.   I don't know. Maybe there'll be no Winchester Street. Oh, I hope. . .

G.   You hope?

E.   Oh, I hope so, for the sake of the others that are coming in after us. They're getting in better places. They're leaving this slums.

G.   Slums!

Sometimes she talks of the recent immigrants. Typically, she is more compassionate than many of her peers. "You've got to love everybody, everybody wants to live." However, there is an air of mystification when she talks of the many young vandals:

> Where do they come from these people? I don't know, that do the damage. You or I don't know. It's done when we're either asleep, or they do it—they do things and run. I just think that sometimes it's that dope that they've got. The kids get crazy, and what comes to their mind, they do it. Their eyes are not clear. I think it's that World War II that made the difference. What I'm thankful for is that I haven't got them in my family, anyway.

Beyond the neighborhood the environment is even more ambiguously construed. It comprises a series of disconnected fragments of experienced milieu. Downtown Lanchester, sections of the suburban communities where her relatives reside, the home of her son in Arizona, the vacation resort in Florida, all form elements in a collage. Each fragment has its distinctive character, its own special geographical constitution. Fragments vary in scale from a remembered room to the lineaments of a city. Often there is considerable contextual variation. Thus, Evelyn can orient herself in terms of the downtown Lanchester of the past, incorporating Cadnum's, Fletcher's, Londergan's, indeed, most of the old now defunct stores. "I thought it was pretty." This awareness of place contrasts with more contemporary constructions of the same space:

> Since they've put those high, them high things up, I think it's changed a whole lot. I think they're trying to make the center big, but it spoils the look of the other parts.
>
> I find it tiresome. You know, I would rather have it the old way, like what we had. Cadnum's, and go from one store to another. What do they have? Those stairs there, those creeping stairs. The stores are so far from one another; but for the young people, they go jolcing around there. I think it's tiresome.

The multiple images are subtly integrated; egocentrically framed, they provide an overall sense of the environment. It is this overall sense, this integration of the past and present of a space which, for Evelyn, identifies the character of the fragment which is downtown.

In sum, fragments of spatially and temporally "beyond spaces," in conjunction with a subtle differentiation of more proximate zones, constitute a complex, malleable, and highly variegated landscape in which Evelyn lives out her life. It is a familiar cozy landscape, the mirror of her life experience. To abandon these spaces would be, in a sense, to lose her identity.

"WE ALL HAVE DREAMS WE'LL SAY"

Evelyn projects an aura of quiet contentment. "God is mighty good to me; that's the way I put it." Would she have lived differently if she had her life over again?

E. I don't think so. Knowing what I know now maybe I wouldn't have come down here—I'd go more to the country. I would have liked the country. A lot of young people are going to the country.

G. But would you have done it then?

E. No. Cause we were near school. We thought everything was handy. School here, and school at St. Mary's, and church. You would sort of think of those things; make it easier on your children, not to go so far in the bad weather and all that.

Gently, without malice, she sometimes rues the passing of a social order which is disintegrating with the demise of the neighborhood. "I think that people were prouder than what they are now, in a way, you know." Occasionally she reveals a sense of abandonment and growing isolation. We were looking through the Christmas cards she had received:

> I've come down. I think I've got only fifty-five. I used to get over a hundred when my husband was living. Oh, last year it came down to about seventy-eight, and this year I'm fifty-five.

Winchester Street is still Evelyn's home. There are many reasons why she would be loath to leave:

> I'd love to go to Arizona if the other two were with me. I'd be with my youngest boy. But then I'd have two left back here. And then, I'll tell you why. My husband has done things here. It's awful to talk about; we all expect to die anyway, sometime. We have a cemetery lot. So if something happened way down there, they'd have to move me way back here. That would be one reason. Then I have my sisters and brothers.

There is also her affinity for the church and the French community:

> You know a lot of them there, that go, have been like from the start, you know, the beginning of the church. I mean we've been here in the parish so many years that it's home. I think it's just natural that we've been so many years and we hate to get away from here. That's about it.

However, there is the characteristic pragmatism:

> Well, like I said, my health is all right. Say something should happen, then maybe I would have to make a move on that account. You never know. Cause I don't know what the future brings for me.

So Evelyn leads an unspectacular life. She is fulfilled, at one within her milieu. There is no reason for disenchantment. I once asked her what her aspirations had been as a young girl. "What were your dreams?" I asked. "Oh, having a family and owning a home. Things like that."

CHAPTER VI

EDWARD

"You adjust to your environment: you have to."

This chapter is written from Edward's vantage point. The dialogue and much of the text is transcribed directly from tapes of our interactions. Some grammatical changes have been made to preserve the continuity of the writing. The material was verified, and modifications suggested by Edward, in a series of meetings at the farm where he spent the last few weeks of his life. During these sessions I read him the various drafts and we worked together on refining the manuscript. Consequently, the chapter is a product of interpersonal dialogue, a mutual creation rather than solely the imposed interpretation of an outsider. For this I am indebted to Edward.

MEETING

I first heard about Graham via the East Lanchester Neighborhood Center Newsletter. My sister-in-law, Francine, who lived downstairs, read the column to me:

> . . .He is anxious to meet with residents of the area (60 years of age and above) who have lived in the neighborhood for many years. If you live on one of the streets listed below and would like to talk with him, please contact the center.

Perhaps I would be interested? After all, hadn't I lived in the neighborhood for all but ten of my seventy-eight years? I was mildly curious, so I suggested she contact the center. A meeting was duly arranged.

I was in the garden standing by the gate when he arrived, accompanied by an older woman. He wasn't very tall and had a dark beard but, because of my eyes, I could not make out his features. Greetings were exchanged and I ushered him inside. His companion, Louise, a caseworker it turned out, soon became engrossed in conversation with my sister-in-law. They drifted into the other room. We were alone apart from the background murmurings of their voices.

Graham started to explain his project. It was something about the neighborhood in the past, elderly residents' memories, the meaning of space, growing old, relationships with environment. His eagerness barely

127

concealed an air of pleading as he sought my help. Would I be interested in participating? He seemed a little confused but relatively harmless. I agreed to help even though he proposed more meetings than I had anticipated.

He switched on his tape recorder and asked me to talk a little about myself and my early life. Sensing he wanted me to start at the beginning, I began:

E. I was born over on Maxwell Street (less than two hundred yards from where we were sitting). I think it was 1873 that this street was opened up for house lots. I heard my mother say that my grandfather had his choice of the lots because he bought the first one; and he didn't want to be on the corner so he selected the third lot down. That was this one. And he lived here with his family until he died. So when my grandfather died, he left the house to his son, Roger. Roger was unmarried, and he sold the house to my parents, and then he lived in. He lived with us while I was a youngster.

G. So you don't remember living anywhere else?

E. No, I don't remember anywhere but this place here. When we moved as youngsters, this house wasn't built. We lived in the cottage, that's where we lived. This was one lot.

I gestured towards the side window where he could view the small cottage which had been our original home.

I told him about the large apple tree in the center of our lot, the three or four pear trees which had surrounded it, the lilac bush up in the corner, and the barn which had been at the back of the lot. We also had a long bench; seven or eight people could sit on it. It was up near the lilac bush. We had several currant bushes in the corners of the lot. We also had tiger lilies that my grandfather planted in the eighteen nineties. They are still coming up in various parts of the yard. And I mentioned the goat. We had received it as a present from my uncle. We got a wagon that went with it; it would seat five of us. There's a picture of the five of us in the wagon. We had that wagon for quite a few years.

I explained how the three-deckers had been built between the cottages, the reason why the house we were sitting in was numbered 73½. My father had built the house in 1911. Well, you see, we had five children, actually we had six. My mother had six, but one died. So there were five children, my uncle, and my father and mother, that's eight. We lived in there, in that tiny cottage. I don't see how. But anyway, in 1911 my father built the house. I think it cost him around

128

$5000. He worked at Restic and Fulton where he made about twenty-five dollars a week. That was a lot of money before World War I. We had enough money. We weren't wealthy but we had enough money. I can remember my father asking me to watch the builders. They were supposed to put tar paper under the clapboard, but if you didn't watch them they put a cheaper material. So I had to watch them to make sure they put the good stuff on.

Graham asked what the neighborhood had been like when I was a child:

E. At first there was only a sidewalk on this side.

G. Was the street paved?

E. Not in the early days, not before, probably just before World War I. Our street was muddy in the spring, rutty in winter, and a sidewalk on one side, this side. We used to play ball in the street, and you didn't have to worry about anything because the horse's head was almost over your neck before you had to move. This was before the automobile came into use.

I remember describing the stores and some of the people who had lived on Imperial Street. We had been wary of the area from there on down as far as the railroad on Eastway Street. Coming home at night, if we were coming home from the city at night, it was scary. There was only the mill and a big long house; the rest of it was open. So we'd rush through there. We'd run. Especially the night I went to see 'Dr. Jekyll and Mr. Hyde.' Boy I was scared! Nothing ever happened. It was just the thought of it.

Many of our neighbors worked in Thomas's or Restic and Fulton. Others had jobs in the local stores; and then, there were a few letter carriers, a few policemen and a few firemen. Most of my friends lived in the vicinity. In those days, my crowd, we chummed around with those that lived around our section. Now Hill Street, up there, well it was a little better area in those days because it was part of the other side of the tracks, the railroad tracks. They formed a barrier. All the people on this side of the tracks went to Winchester Street School, but most of them on the other side went to Forest Street School, which was considered a good school—a better school than Winchester, that is, the type of people. They called it the "West Side," Oakley Street and Forest Street. We knew them, but we knew most on this side of the tracks.

As I explained this to Graham I could sense he was beginning to get a false impression of a purely local childhood. So I told him about some of my activities beyond the neighborhood. I remember being interested in politics at an early age. In nineteen hundred and eight my friend and I wanted to find out about the elections. Taft was running against Bryan. You couldn't get the report down on Andrews Street. My friend and I were perhaps twelve or thirteen. We went downtown and we went into the theater and sat up in the gallery because we knew they'd give a report sometime during the show. It was a musical play, I can remember to this day. And the girl came out and sang, "William Howard Taft is the star I'd like to be." And the next time she'd come out she'd sing, "William Jennings Bryan is the star I'd like to be." Anyway, near the end of the show, the manager came out and he said, "We have an announcement to make, that William Howard Taft is elected President of the United States." The same woman then came out and sang "William Howard Taft is the star I'd like to be" again. And of course I was happy then, because my father was dyed-in-the-wool Republican. My mother was a Democrat!

I also described some of the excursions we would take. You see, it was all trolleys in those days. In the summer many people would get on the trolley and go down Lake Quaninon for five cents. Sunday afternoons, the most popular place was Lake Quaninon. There was also another place, out on York Street, called Firecrest, which was an amusement park. I can remember my friend and I one Sunday deciding to visit every park in the city—to take them one at a time. I also remember taking the trolley to Boston, before automobiles were common, and getting on the boat and going to Gloucester. Of course, you'd come back the same day. That would be as far as you'd go. Although before World War I, when my father had a week off in the summer, we would sometimes stay with his brother in Bridgeport, Connecticut. His brother would always take us to New York. That's when I got my first look at the Statue of Liberty and Grant's Tomb. My aunt always said that I was a good one to entertain because no matter what they suggested, I always said yes.

Then there was the time I went to New York with my friend. We'd be about sixteen or seventeen. We left Lanchester and we got on the night boat at Providence, and went from Providence to New York. We got in at about six o'clock in the morning and stayed in the St. James Hotel. We found out that you could take boat rides. We'd take a ferry and go over to Newark, New Jersey. We'd send a postcard to somebody from Newark. Then we'd go back to New York and we'd take a ferry over to Weehawken, and later on, to Jersey City. In each one of these towns we'd send postcards. Then we finally took the Albany boat, as we wanted to go as far up as we could go and come back the same day. So we went to Newbury. That was quite a trip!

Conversation turned to discussion of my education. In those days very few from our segment went to high school. Most of the people would go to Winchester Street, and then they'd just go to work after that. My best friend, Joe Gardner, he went to work in the mill, like most of my friends. But my mother was always interested in having us get an education, so we went to high school. My young brother, he didn't want to go, he only went two years to the high school, but the rest of us went through high school. I remember getting acquainted with fellows and girls who had more money than we did. Well, I got away from most of the neighborhood fellows when I went to school; not so much in high school, but when I went to college I really got away. Sometimes I wouldn't see any of them for six weeks. Eventually you'd kind of drift away.

In high school I was a very quiet boy. Even now I speak softly. I would blush easily. I remember the high school teachers were trying to get me to be a little more forward. So suddenly, out of a clear blue sky I was called down to the office and assigned to deliver messages from one room to another. Well, on one occasion, I was a sophomore, I walked into the senior room and handed the teacher a note. And as I turned I saw this neighbor of mine, Claudette Rousseau her name was, a good-looking girl, she was two or three years older than I was. She looked up at me and smiled and winked, and without giving it a thought I just winked back. Then of course the whole class saw me. Nobody saw her. I must have turned pink. They all started shuffling their feet and clapping their hands. When the school paper came out it wanted to know who the senior was that Edward Foster the sophomore was winking at in Room 11!

Anyway, the teachers liked me and they helped me to get into Marsden University, here in the city. I entered in the fall of 1913 and graduated in June, 1916. During the summer vacations I worked one summer in Restic and Fulton, one summer in Thomas's, and one summer I carried milk. Then I went to graduate school and completed my master's degree in history in 1917.

I paused to check that this was the kind of information Graham wanted. "Yes, yes," he replied, "tell me what you did then." So I told him how, after a further year in graduate school and six months in the Army, I became a teacher. I spent a year at Pelton High School, and then, between 1920 and 1928, I taught in Mintern. This was the only period of my life when I did not live on Andrews Street apart from the

131

two years after I was born. In the summer of 1928 I returned to Lanchester and taught at Collaton High School until 1951 when, at fifty-five, I took an early retirement.

Oh, I missed the kids. I missed the youngsters, but not the teachers and not the principals. Most of the principals were political hacks! After retiring I was able to follow the stock market more closely, it had always been one of my hobbies; but I haven't worked since 1951. I never married although I've always had girlfriends.

By this time over an hour and a half had passed. Louise had come back into the room. She seemed anxious to leave, and I was getting a little tired. I told Graham I felt we should only meet for an hour or so at a time. We scheduled a second meeting for the following week. As he packed up the tape recorder and put it in his case he kept thanking me.

After he had gone I sat and pondered for a while. I still wasn't sure what he was trying to accomplish. But then, if I could help. . . I wondered what would happen when he called the following week.

LIFESTYLE

During out next few meetings Graham seemed more interested in my contemporary situation. I told him about some of the limitations on my lifestyle. The situation at home was one of them.

I shared the third floor apartment with my sister who was partially paralyzed. She was one of those who wouldn't accept that she was paralyzed. She wouldn't do therapy because, at first she could get up and down the stairs, with a little help, and she could walk to the kitchen. When they don't accept it, well, there isn't much you can do. It was not a good setup. She was always the tyrannical bossy type even when my father was alive. She always had her own way.

By this time we had a housekeeper. She was no cook at all: everything was fried, she never put anything in the oven. You didn't starve to death, but you didn't eat like a prince. No, the meals weren't what they should have been. She didn't believe in vegetables; you had to ask for vegetables to get them. For greens I got a pepper pretty nearly every day. She couldn't get it through her head that I didn't want peppers. We had to eat to satisfy the housekeeper. I'd get my own breakfast when I got up. The housekeeper would come in at eight; she'd get my sister's breakfast at eight. Dinner would be at about twenty-five minutes of eleven, that's when it came on the table, that's when I'd have my pepper! We were all through at eleven. Then at two thirty she'd take my sister out to the kitchen and give her lunch. So the housekeeper could go home at three. I'd get my own lunch. Sometimes I'd just have a piece of cheese on toast. And I'd have half a banana or something like that. I'd probably eat around four, four thirty. At night my sister might have a glass of water and a bran cracker before she went to bed. The housekeeper would come over and put her to bed at seven thirty. It was not a good setup.

A second problem was that I could no longer drive. I remember my first automobile ride. It was out at Washington Park. My friend and I took a trolley ride out there and, coming back, a fellow was coming along in a car. I had never gone in a car. I motioned to him and he stopped. He must have been in his twenties—nice fellow—and we were kids. He picked us up and drove us to the city. We used to laugh at the first cars. Nine times out of ten they'd come back being driven by a horse. It was a common expression, "Why don't you get a horse?!" Well, anyway, I bought a Buick in 1928; it cost me $1200. I think the car tended to make people less sociable. Before we had the car we knew everybody. When you were out walking you ran into everybody. Once you had a car you no longer mingled with the fellow next door. You didn't even see him. You'd come out of your house and get in your car and take off. I'd get out a lot when I had my car. I quit driving at the end of '65 because I knew that my eyes were bad enough for me to get in trouble if I kept on. I could have driven probably another year but I figured it was dangerous. I haven't been driving now for ten years. You miss a car. You get used to it. You miss the ability to take off when you feel like it. If I get up now and it's a nice day I can't go for a drive. In the past if I got tired of reading I'd say, "Oh gee, I guess I'll take a ride out in the car," or, "I guess I'll take a ride downtown." But I can't do that any more.

I am legally blind but if I take my glasses off nobody knows. I can tell light and dark. The doctor said I'd always be able to tell light and dark. He said it would be a little awkward, and I know now what he meant. If someone spoke to me on the street or waved to me from a car I wouldn't see them. If you walk in a store where you don't know where everything is, you have a hard time discriminating among the articles. Some of the old stores were all right because I knew everything there. If I wanted candy, I knew where they sold the candy; if I wanted a book, I knew where the books were; if I wanted perfume, or if I were buying a birthday present, I knew where that was. When I'm out in a car, the people I'm with, they all know I know the roads, but the minute we hit a new highway I'm lost. If it's an old highway, I can tell them where to go, but on the new ones I can't.

The blindness has not really limited my walking. I can still step out briskly or I can stroll around. I pay a little more attention and I'm more careful where I put my feet down, but it doesn't slow me up although I have to be careful crossing streets. I can see a certain distance so that I can tell whether a car is coming. It's where the traffic is fast that you have to be more careful.

You develop certain skills when you're in my position. With traffic lights, when it says "Walk—Don't Walk," I can't see it. There's where you have to learn. At the traffic lights at the end of my street, if there was no traffic coming up Andrews Street I would press the button. I could see the yellow light come on, and if cars were coming along Durham they'd stop. Now if cars were coming up Andrews Street, I wouldn't press the button. I'd wait for the cars because practically all of them turn left and go down Durham. So I'd wait for the first car, and if it went down Durham Street I'd cross. I built a system for dealing with little things like that. Then you watch pedestrians. If I come to a corner and there are people standing there, when I want to cross, I just wait for them—watch them. When they move, I move.

You tend to stick to the same routes. You go to buildings that you know pretty well. When my doctor died, I picked another one in the same building because I have difficulty finding numbers if I go to a strange building. When I go shopping, often I'll have somebody with me now. Sometimes I'll go alone, but I will probably ask a clerk or floorman for help. I put one dollar bills in my right pocket and tens in my left. I try not to have any fives or twenties.

Of course you learn one thing, that people are very nice. I remember being in the A & P. I simply said to this woman, she was standing there looking at the shelf, "My eyes are not too good. I wonder if you could tell me if such and such a polish is there?" And she said, "Yes. Do you want the liquid or the powder?" You learn you have to do those things, but I was never embarrassed to talk to people. I can talk to anybody. Of course, when you talk to people you're a little more sensitive to the impression you might make. You don't say too much until you know who they are. You're a little more careful.

My lifestyle was adjusted to these constraints. I'd spend much of my day, about four or five hours, reading. On a wet day I'd read most of the time. I don't kill time, but the way I would get rid of time when I was home was to do a lot of reading. As I could not see, I'd get books on records from the library. They would keep me so that I never ran out. People who can't read, I sometimes wonder how they pass the time, whereas I can sit down in the chair, light up my pipe, and stay there for three or four hours.

I also walked a lot. It seemed as though I could walk as long as I wanted to. I'd walk around the neighborhood, all over, sometimes down as far as the end of the mill. I prefer to be out where I don't have to watch traffic, so I particularly liked walking in the cemetery. I'd wander for hours in there. Sometimes I would just walk up and down the street. It wasn't dangerous right around our house, although in other sections close by, particularly as you got up to Imperial Street, they sometimes had trouble. I couldn't see too well, but the place looked as though it was run down and there seemed to be the rougher element there. I wasn't frightened of walking, although I would never go out at night—not alone. If I were downtown and had to get back at night, I'd take a cab home. No. I didn't go out after dark. See, you were not sure if it was dangerous. I remember one time this fellow said to me:

> If I was coming up the road and you were coming down towards me and there was nobody else around, we'd both be frightened. If one made a dash for the other, the other would turn round and skip.

Either one would scoot if the other made a move!

I was pretty independent. If there was anything I needed, or if I wanted a ride, I could call upon my nephew or upon my niece. I wasn't imposing because they knew I didn't like to do it too often and I always made sure that it didn't cost them.

I'd get away from the neighborhood every chance I could get. I would travel downtown, usually once a week, to have dinner at one of the restaurants, the Diner's Haven. Well, I knew the owner and the waitresses. One of them was one of my old pupils. She knew me very well, and when I'd come in she'd take care of me, she'd find a booth for me. Downtown were all the people that I'd known years before, but I couldn't recognize them. A woman stopped me a while ago. She said, "Oh Ted!" If she hadn't told me about her husband dying, I'd never have known who she was! Formerly, I could go downtown and just by running into fellows I knew, I could spend a couple of hours just talking for five or ten minutes. But by this time, if somebody went by I couldn't tell whether I knew them or not; they'd have to recognize me first. People I hadn't seen for a few years wouldn't stop to figure that I couldn't see. It bothered me a little bit to know that there were people going past that I knew.

On many Saturdays I'd come out here to the farm. My nephew would drive me out and I'd take the two dogs for a walk. I'd usually walk for about four hours. I could roam around here, whereas I felt sort of confined in the city. Out here you didn't have to worry about traffic, just keep to the right of the road. You'd only see about three or four cars. When I was younger I'd come out to the farm one or twice a week. My cousin's children who now run the farm were kids then. They remembered me, so I could come out here any time. And they would take me sometimes down to Boston. They were apt to call me any time during the week and ask me if I wanted to take a ride to Boston, and I'd always say yes. So they'd take me to Boston, and when we were done, we'd go to some restaurant. I can also remember taking trips to Canada, New York State, and Pennsylvania with them.

During the summer, since I couldn't drive any more, I'd go up to Hampton Beach. I'd stay there all summer. My nephew would drive me, and when I got up there I'd meet people that owned cars. They found me rather interesting, at least that's what they said, so they used to take me out riding to Maine and New Hampshire. They'd take pretty good care of me up there. This friend of mine had a nice automobile, a Monte Carlo. Well, it was a woman I met up there the summer before last. I met her in a group. She was talking—we got talking—she asked me to come up and see her when I was up her way. She told me where she lived. So I dropped in, and from there on, why, any time I wanted to go out, we'd go out. We've eaten in some pretty nice restaurants.

You see, I got around a little. I never was tied down.

THE PAST OF PLACES PRESENT

After we had been meeting for several weeks it was agreed that I would show Graham around the neighborhood. It turned out to be a clear day, sunny but not too hot. I was on the pavement outside when the car pulled up. He helped me in. I heard the tape recorder click on as we moved away from the curb. We traveled only a few yards before stopping at the lights. I could hear the tick of his indicator. He didn't say anything.

"The same person owned this whole block," I offered. He seemed not to hear. "What about Killarney's Park, was that there?" He was obviously looking out over the small park in front of us. There was a baseball diamond and asphalt basketball court on the far side. "Yes, but it wasn't owned by the city," I replied.

It was owned by the Struthers and Anson shop. There was a big house on the corner that's been pulled down now. There's a handball court there now. Can you see it?

There was a rustle of his shirt as I sensed his affirmative nod. I continued:

Old man Killarney lived farther down the street. Right over there was Killarney's house. That's the community house for the East Lanchester Playground now, but when I was a child, Killarney's house was there. He was a Civil War veteran and also a member of the police force. About where the swimming pool is, a street ran through there called Sentry Court, and there were three or four houses on that street.

At last the lights changed. We turned left. I could sense our progress along Durham, over the bridge, and down towards the "Four Corners." I explained that on the left, where the Welford Nursing Home now stood, had been the David P. Caxton estate. This was where as youngsters we used to steal apples. They had a lot of apple trees. When we reached the "Four Corners," I described what had been at the crossroads:

This, on the right here, that was Thomas's. It was on the hill, see. Made a beautiful sight.

I could picture the mansion which once stood opposite St. Mark's Episcopal Church, which old man Thomas had also commissioned and financed. The mansion had been demolished to make way for a supermarket with its large parking lot. The church remained. I told him about the pharmacy which still survived on the corner to our left:

That's where we used to come for our prescriptions, and ice cream. In those days the pharmacies carried, all pharmacies had a soda fountain. And we used to get ice cream sodas.

As we turned onto Eastway Street I described where a grocery store had been, the barber's shop, the liquor store, Porter's Hardware Store, Vale's, the store where we used to get our shoes (his son was later Governor of New Jersey), and Faith Church. Only the pharmacy and the church were here now.

I told him to turn left up Maxwell Street because I wanted him to see my birthplace, a three-decker on the right-hand side. When we reached the top of the hill, the street was unpaved. We bumped along between the dilapidated houses: this area had seen better days. We came to a gap between the tenements and, at my suggestion, left the car and walked to a small rise which was littered with concrete rubble and beer cans. From here he could look out across the railroad tracks and down over the neighborhood beyond from which we had just traveled. I could imagine the scene before his gaze. It wasn't an attractive view these days.

Over the tracks the land had belonged to the Paines. They were very unpopular because they wouldn't let anybody play in their field. If we were in the Paines' field, we had to watch out. I explained about the railroad tracks below us:

The railroad track here was level with the street. The railroad's been lowered since. It used to be level with the road. I know when I used to peddle newspapers, sometimes there'd be a delay of fifteen minutes, and you'd take a chance of running under, which was dangerous. Youngsters used to hop the freight, you know. Quite a few of them got killed. That was a common thing in those days, to hop a freight.

"That was equivalent to streaking!" he commented lightly. "Yeah," I replied:

Only more dangerous. Our mothers used to warn us not to hop freights. I've seen people that were killed. I saw a fellow with his leg off one Sunday morning.

"That was down here?" He was more serious now.

Yes. Down on the railroad track. And there was a fellow I knew very well—young fellow. He was running along the top of the freight cars, and he stumbled and went down between the cars. The brakes were made of iron. He was carried away dead: fractured skull. So the railroads were a problem in those days.

"That must have made quite an impression on you," he said quietly.

We turned and headed back to the car. We traveled all over the neighborhood that day, and as we journeyed I talked of the places we passed. I showed him Holy Sacrament Church where I went during my brief enforced encounter with religion, and the house where Father Smith's Temperance Society held their meetings. In those days, before World War I, every church had a temperance society. I took him down past the Thomas's plant and showed him the entrance where I went in the day I got my summer job there. The mill itself was now a furniture warehouse. As we were traveling up Treeside Street I remember he

138

seemed to be going too fast. "Watch out!" I warned, recalling the potholes. When I was driving, it was very bad.

I told him about all the stores. There was a store on every corner. We were on Durham Street passing the end of my street:

E.   There was Shaugnessy's store on this corner. On this side of the corner was a barber's shop, and that's where we used to pick up our newspapers at night. He had a newsroom in the back. Then, if you kept walking on the same side, across the street, you came to McNally's Market.

G.   Was that McNally and Holmes?

E.   McNally at first, then McNally and Holmes, and Holmes later. He hired Holmes as a kid. Holmes would be my age. And then when he died, Holmes took over. Across the street from McNally and Holmes was Hennessey's Market, and Hennessey sold out to Reilly. So it was Reilly's Market later on. Then if you went up a little way from McNally's Market, on this side was Sullivan's store.

We must have driven around for three hours. As we were returning to my home, I suggested we stop at the British Men's Club for a beer. The club is only a few yards from the house. It has been in existence since 1895, the year I was born. It was all British then. I have a key because my family was English. I don't know how many full members they have now, but it must be down around seven or eight because nobody was born in Britain that I know of. I'd go in a few times a year. Sometimes my nephew would visit and he'd want to talk, sort of confidential like. He'd say, "Let's go over to the Club." It's noisier over there these days. They're not really my type. I could sit there and spend half an hour, or three quarters, but not too long. I didn't like to talk baseball all the time.

I think Graham enjoyed the afternoon. So I promised to show him some of the other places I went. I asked him if he'd like me to show him around the cemetery. "Sure," he said. But he seemed a little puzzled.

HOLES IN FENCES

See the iron fence? Well, they didn't have that iron fence until about a year ago. So I used to come over here and come up this street and go through the brush. There'd be an opening in the brush and I'd go through.

139

We stood less then two hundred yards from my home, at the top of the dead end street where I used to enter St. Peter's Cemetery. They had put a fence up, but there was a gap I could squeeze through (Plate VI.1). When I didn't want to walk too far I wouldn't enter the cemetery. There wasn't much traffic and you could walk back and forth here as well as anywhere. On the left, as you faced the cemetery, was Killarney's Park with its swimming pool and community house. (I was playground director in the summers of 1919, 1920, 1921, 1922. We got fifteen dollars a week.) I always knew the playground attendants and sometimes I'd drop in to talk with them. If they didn't see me for a few weeks, they'd wonder if I was sick. The week before, the woman who works there had asked me where I'd been. There was one fellow I talked to a lot, an inspector; he would be inspecting the swimming pool to make sure everything was done right. Of course, he'd just be standing around there most of the time, and he liked to talk.

We walked back down the street towards Graham's car. Then I caught sight of Caesar bounding towards us. I would recognize him anywhere. He leapt up at us, excitedly licking and pawing. All the dogs in the neighborhood knew me. If I went for a walk I'd have at least two and possibly four with me. "He belongs to the Cousins," I explained:

> ...a fellow named Cousins who lives down here. He's so used
> to me coming over the playground. Now if I was going to walk
> and I came up here, and went through the hole, he'd come
> right with me. But as soon as I start that way (I pointed
> towards my house), he'll go back home.

All the dogs in the neighborhood used to wait for me. It had gotten so that I couldn't watch them as closely as I'd like. I decided I'd have to break them from coming down to the house every morning; they would be there at seven o'clock in the morning, just waiting for me. So I started to go out later in the day. I'd probably only run into one of them then. I didn't mind one, but not four, unless I could see well enough to check them.

We got into the car and drove round to the main entrance of the cemetery. As we entered Graham asked me how often I came to this place:

E.   I used to come over here twice a day and just roam around, you know, in the morning and the afternoon. But now, on account of my eyes, I don't do it as often as I used to.

G.   But you still do it sometimes?

Plate VI.1 "There was a gap I could squeeze through"

E. Oh yes. Yes.

G. Doesn't it ever depress you?

E. No. Well, you see, we were brought up so near it. Everybody down my way used to come over here. But people from the other side of the city would say, "Aren't you afraid?" But it was like a playground for us.

I was hoping he'd understand what I meant before the afternoon was over.

When we were near Denlon's grave, I asked him to stop. Denlon of the Lanchester Regiment, he was their general. A captain when they left Lanchester, he then became major and eventually was a general. He was also Chief of Police in Lanchester for quite a few years. Nice fellow. I liked him very much.

E. Now these graves here were moved from Merton. They were buried in this corner for some reason.

G. How did you know that?

E. Well, I could read them,years ago. There's a fellow buried over here that fought in the Crimean War as well as the Civil War. Isn't there a sort of half tree there?

G. Yes, there is.

E. Yes, that's the old one. I think his name was Patrick, Captain Patrick.

There used to be a tombstone over here somewhere, I'm not sure where it is now but they had a chalice on the top of it. You could take the top of the chalice off and there was money in there. The kids found it out.

G. So you must know these graves very well?

E. I used to know where everybody was buried. People just asked me.

The main reason for stopping here was to show Graham my family grave. "My folks are buried in there," I said as we stood by the plot. "The whole family's down there except my sister and me. There's room for three more." I remembered my mother:

When we were youngsters, you know, I remember when I was probably four or five, my mother'd say to one of her friends, she was going to take a walk over to the burial grounds. They used to call it the burial grounds. I used to hate it because the

142

roads weren't paved. Of course, this would be in the summer, and they were dusty, and it was hot and you didn't like it. Even today you'll hear people say that the cemetery is probably the coldest place in the winter and the hottest place in the summer.

When we were youngsters, see, people didn't have cars, so on Memorial Day everybody would come to the cemetery; the place was packed. In fact, there were crowds over here for a week before. If I came over here with a lawnmower or a sickle, somebody might ask me to cut the grass, or they might ask me if they could borrow the lawnmower. You would make four or five dollars in no time. Later the cemetery changed the rule; you'd have to have your lot taken care of by the cemetery. But when I was a youngster we could make a fortune over here.

Graham was silent. At first I thought he was just listening to what I was saying, but gradually I sensed something else. He wasn't used to cemeteries. He seemed to be brooding; for a while he seemed lost within himself. I startled him from his private thoughts by suggesting that we return to the car.

Driving slowly onward, we came to a fence separating St. Peter's from a small Jewish cemetery. Mr. Thomas, English founder of the carpet mill, had bought a field up here. They called it the "cricket field." I saw my first and only cricket game here. The English fellows played, and they played soccer and baseball, too. To me it will always be Thomas's cricket field.

Leaving the old part of the cemetery, we crossed the bridge over the Severdon River which passes through the cemetery, and entered the more recently developed area beyond. After a short distance I asked Graham to stop and we left the car. "There were some beautiful trees right in here," I said. As we walked across the grass, I told him about the shady maple grove where the hoboes used to hang out. They had wooden boxes then. If there was going to be a funeral, the wooden box would be on the top of the grave the night before. So these fellows would get drunk and lie down in the box, and pull the cover up enough so that they could keep warm. More than one fellow going down Durham Street in the morning saw somebody getting out of the box!

We reached the railings which formed a barrier between St. Peter's Cemetery and Faith Cemetery. Faith Cemetery was the city graveyard. Beyond the fence the ground slopes up a small hill. Some people called it "society hill" because all the wealthy people were buried on the top.

There were many impressive monuments. In the corner, the railings were bent out of shape (Plate VI.2).

E. I go through here, see, at least I used to go through here until some kids knocked over a lot of the tombstones, until they closed it up. And I'd go through Faith at least five times a week.

G. Oh, I see; all this barbed wire.

E. We used to get through here.

G. Oh, I can see, in the corner.

E. I can still get through.

G. You can?

E. If you get down low enough, yes.

We went back to the car and continued onward, passing an area where many of my friends and former colleagues are buried. Albert Lepellier, John Roberts, Charles Synon, Joe Doherty, and others, they're all in there now. The headstones are pretty well spaced in that part.

Next, we stopped on the far side of the cemetery. This area was once swampy woodland. There used to be a beautiful natural spring under a big tree, where we would sometimes pause for a drink. The land was filled in the years before the interstate highway, which stood in front of us, was driven through. Now there's a lot of turf. Where we stood there were a few graves to our left. On the right, where the ground sloped up into Faith Cemetery, was where the "jungle" used to be. It was like a jungle. When I came home from school, I could take off and go through the cemetery, and within fifteen minutes I would be in the woods. In the fall you could pick a pocket full of chestnuts, the little ones, within fifteen minutes. Of course, those trees have all died of disease since that time, but everybody loved them. My mother liked the chestnuts. On Sundays, when the boys couldn't play baseball, quite often they'd go out into the woods and they'd have a crap game. They'd keep their eyes open for the police, but very few arrests were ever made because the police were more interested in getting them to run and leave the money, so that they could pick up the money. Sometimes we'd pass right through the woods to go swimming at a pond beyond where the highway now runs.

Graham seemed interested in the highway. I explained what it meant to me:

144

Plate VI.2 "I can still get through"

145

In the past few years they've been closing in on me. The first thing they did was to put that road in. And then the next thing they did was to start setting the graves up here on the left. So they've been gradually cutting off the places where you could roam without being in the cemetery.

We started to drive back towards the bridge. We passed through the area which had been Brent's Farm. He had a couple of fields in that section of the cemetery. The bridge was not in existence when we were kids. The first time I crossed I crawled over a big pine tree, a pine log that straddled the stream. Eventually, of course, you got the knack of walking over it. It was marshy here, too. By the river, the woods were loaded with snakes and frogs. You could get frogs' legs if you liked frogs' legs. I used to go over sometimes to kill snakes, and my brother would go over to catch frogs or shoot squirrels. He used to like to hunt chipmunks with an air rifle. I didn't care for shooting and hunting.

There were no headstones on the graves in this area. Much of the land was fill. The graves had a tendency to sink, so the memorials had to be flat. If they put monuments up, they'd have gone out of sight, or at least tilted badly. A lot of the people that had those lots didn't realize that the ground they were in was just fill—poor fill, too.

Our last stop, on the way out, was back in the oldest part of the cemetery, near the shed where I used to sometimes sit and talk with the workers. We were friendly enough although the dogs which usually accompanied me were not supposed to be in the cemetery. I'd tell the workers:

Stray dogs run over here, rabbits are over here. What difference does it make? Especially if the dog is with you.

I'd say, "They're not my dogs." I could say that honestly, even though I'd have three of them with me, they'd be following me. We'd laugh.

RAINY DAY

On the week following the trip to the cemetery I had planned to bring Graham out to the farm, but it was raining when he arrived so we postponed the visit. Instead, we talked about my friends. I was always the friendly type, most of the dogs knew that! In a way, the dogs were my closest friends. Most people who know me would say that. You see, I couldn't possibly keep close to all the people that I knew, there were too many of them. There were many that I could stop and talk with, but to say that I was a close friend? If they heard about it they'd wonder what I meant. No, I had many good friends, but not close friends.

Some of the people I'd known on the street had died, and others had moved, but there were one or two who still lived around. I'd meet them when I was out walking; we'd greet each other. At the time of this conversation, I remember, I'd just met Bill Granger. He had been going by and saw me in the yard, so he spoke. I had got up and gone to the gate and chatted with him for half an hour. He used to roam around too. After a few minutes Graham produced a list of names of elderly people who lived in the neighborhood. He asked me to tell him if I recognized any of the names, and then he started to read them to me. "Edward Foster? Never heard of him!" I said when he came to my name. I knew mainly the old ones. Many I'd never heard of (Figure VI.1), although I might have known some of them before they married and changed names. The housekeeper upstairs would probably have known them all, especially the French ones. I knew very few of the French people.

Apart from those I'd known when I was a youngster, the people up at the pool, and one or two others, most of my friends came from outside the section. I suppose this was related to my education. Most of those in the district, of my generation, only went to sixth and seventh grade. They were nice people, likable. But sometimes their conversation was dull because they had nothing really to talk about. All they could talk about was people. I was more friendly with those I'd known in college, people who had been colleagues or pupils when I was teaching, and friends I'd met when I was in the American Legion. (I was commander of Post 7 at one time.) I knew many people around town. I was well known in the city. I taught people who were presidents of banks, nurses, company officials, policemen, all kinds. If I went into a lawyer's office or a doctor's office, if I went into a hospital, if I went into the police station, I always met old pupils. I'd always get good service. And, of course, there were the people I knew here on the farm and those I met at Hampton Beach.

As you get older most of your friends die. Some that I knew in the forties, and a good many I knew in the thirties, are dead. We talked about dying and the loss of friends:

E. Now I can't tell if they die. If they die now, I don't even know about it. I can't read the paper, the obituary column.

G. Did you used to read the obituary column?

E. Oh yes, always, as I got older. I think older people do.

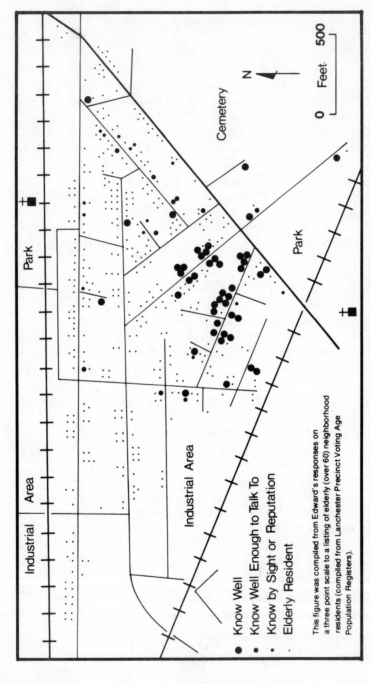

Figure VI.1 Edward's Winchester Street Peer Group Social Network

Know Well

Know Well Enough to Talk To

Know by Sight or Reputation

Elderly Resident

This figure was compiled from Edward's responses on
a three point scale to a listing of elderly (over 60) neighborhood
residents (compiled from Lanchester Precinct Voting Age
Population Registers).

Cemetery

Park

Park

Industrial Area

Industrial   Area

N

0   Feet: 500

G. That's a thing I've noticed. Doesn't it make them upset?

E. No. I don't think so. I just take it as a matter of course, you know. My sister, in one of her rare moments the other day, asked me if I knew that somebody had died. And I said, "No, I didn't know." I said, "He graduated from high school with me. I knew his brother Bill and his brother Jim. They were all nice fellows, they were all good friends of mine." I don't know how my sister knew that I knew him, unless he belonged to Post 7 in the Legion.

G. So a lot of people that you knew, that as far as you know are still alive, might in fact not be living?

E. That's right. Yes, because some of them would be in their eighties, and some in their late seventies, see.

G. Then you've got no way of keeping up?

E. No.

G. Would you like to?

E. No. It doesn't bother me much. I very seldom. . , I never like the idea of going to a wake. For some reason or other I never liked it. I very seldom went unless I had to. I told my nephew that if anything should happen to me, that I want it to be private.

G. Why?

E. Well, if I didn't like it myself when I was alive, I wouldn't want other people coming in. Well, it puts quite a strain on a lot of people.

OUT TO THE FARM

The next time Graham came the weather was better, so we came out to the farm. It was a familiar twenty-two miles. As we traveled, we talked.

I told him how, since the twenties, I would come here about once a week, sometimes more often. In fact, when I was driving I might come any time I felt like taking a ride. My cousin, Richard Collins, owned the farm; there would be anywhere from fifteen to twenty thousand chickens. Sometimes I'd help with grading eggs, but I didn't help too often. I left that to people who knew what they were doing. There were two children. I was driving then, of course, so I'd take the youngsters out. The children are grown up now. The boy, Richard Jr., he's in his forties, he took over running the farm. Nine years ago they got rid of the chickens and turned it into a dairy farm.

The family also had a camp at Lake Gaugahana. I could go up there, as I had a key. Sunday mornings sometimes, especially on a rainy day, I'd take a ride out to the camp. I'd take the New York Times with me and spend a few hours there reading. When I wanted dinner, I would either come over to the farm or, if they knew I was there, they would call me.

Since the trouble with my eyes, I'd come here less often but for longer periods of time. I remember describing my routine to Graham. I'd get up between six thirty and a quarter to seven. The coffee would already be boiling as Richard gets up at six and puts it on right away. I'd come downstairs with the dogs and have breakfast. I'd have a cup of coffee, cereal, and half a banana, and then I'd boil a couple of eggs and make some toast. I took care of myself! Then I'd take the two dogs and go out.

I'd take a long walk through the woods, or go down to one of the small pastures where the dogs could chase rabbits. I probably wouldn't come back until ten thirty. Then I'd have coffee in the house. I might watch television for a little while. I like the news. I can see it but it's blurry. If the weather was nice, I'd have my chair out on the lawn and sit with the two dogs around me. We'd have dinner at about twelve thirty and everybody would sit around until about one thirty. Then I would usually pick up the mail for them, as the mailman would leave the mail down the drive. After this I would either start out again with the dogs or sit out here on the grass. On rainy days I'd put my chair on the porch. Sometimes, during the winter, if the weather was a little raw, I'd find one of the old cars that they don't use, and I'd sit in whatever seat the sun was coming in for an hour or two. I'd rest, just rest. I'd be outside most of the time. I'd kill time that way. If I was out of doors, I liked it.

I remember Graham picking up on my reference to killing time. I told him I didn't want to just sit down and do nothing. If I was going to be out there for any length of time, I would probably bring my record player along with me, and my records, so that when they had a really bad day I could read. I tried to avoid coming on bad days. I'd watch the weather and if I thought it was going to rain a lot, I wouldn't come. I always tried to figure out whether I'd be a burden to them. I gave that a lot of thought. They knew that when I showed up they didn't have to worry about taking care of me; I'd take care of myself.

150

When we arrived here at the farm, the dogs rushed to greet us. Usually the first thing I'd do would be to take them for a four or five mile walk. But not this day. I showed Graham the broken-down chicken houses and the milk house in the old barn. I also showed him the trench silo. One end was open, but the rest of the corn was protected by a plastic blanket which was held down by worn tires. And I showed him the dump. It was an area where they left the broken machines, the rusty cars, old tires and lumber. There was lots of other rubbish there, too. He seemed fascinated by it all.

After we had finished looking around, I took him into the house. I introduced him to Richard's wife and the two boys. The boys both work on the farm. The sixteen-year-old can do anything on the farm. The twenty-three-year-old, he can too, but he had a bad motorcycle accident and has been in poor condition ever since. He was just beginning to pull out of it at the time, but there was some brain damage.

On the way back to Lanchester we stopped at Aldea's, a restaurant I sometimes visit. Over sandwiches and coffee we talked about the people on the farm—mainly about Ralph. He was very down after the accident because his girlfriend deserted him. While he was in hospital, she started running around with two other fellows. He took it quite hard. I told Graham how one day I talked with him. I said to Ralph:

> I don't think you ought to show bitterness, because you don't want her to think you're jealous because she's going out with somebody else. I used to act as though I didn't care.

I think he was surprised.

> I've never mentioned it to anybody, and I didn't think anybody else knew. If I could only talk about it. Nobody ever hears me say anything about any of the girls I used to go with.

He was worried about the effect of his accident on his social prospects. So I talked it over with him. I told him about some of my own experiences. Right afterwards he left the room, and I went into the kitchen where his mother was working. She looked up and smiled. "I got the message, Ted," she said. She knew what I was trying to do: to get him to accept it, and not talk about it, and just act as though it didn't bother him at all. That would have more effect on the girl than the way he was acting.

Graham and I talked for a long time, we didn't hurry. As we were leaving, I remember his surprise when I told him to mind the small step at the entrance to the foyer. He didn't realize that in my situation you remember those things!

151

REFLECTIONS

Soon after this trip to the farm Graham stopped visiting me. One day he called to say he was unwell and couldn't come for a while. I didn't hear from him for several months. I took a trip to Hampton Beach and spent a lot of time on the farm. In November I became very ill and was admitted to Mercy Hospital where I had several stomach operations. I couldn't take solid food for almost three months. Shortly after Christmas, Richard, on one of his visits, told me I had received a Christmas card from Graham. I wondered if he had been trying to contact me, and I asked Richard to call him and tell him of my situation.

"Mr. Foster? This is Graham." It was three days later, the day before I took my first solid food. We talked a bit, but I tired rapidly so he didn't stay for long. On later visits he told me that he had been writing about me and that, as we had originally agreed, he would like me to help him with the drafts. He said he could come to the farm for this purpose, for by this time I had decided not to return to Andrews Street and, instead, to live here permanently.

Sitting here on the lawn with the two dogs at my feet, or when I'm out walking, I often think of things that happened to me. I reflect on my years at Collaton High. I recall the pupils. I recall fellows I was with in the army, and fellows I went to college with. And sometimes I think about places I went. I'm the type of person that likes to dream a little. Dream, I call it dreaming. I think it's good for you; just takes up your time. I try to keep off sickness and things that bother me, but I'll often just sit and think about the things that happened. I have a tendency to think of what pleases me whereas some people might think of things that make them bitter. I don't have any regrets. I don't look back negatively. Most of my thoughts are on pleasant things. Oh yes. Sure, I often sit and dream.

Occasionally, I reflect on my times with Graham. In particular, I recall conversations about my feelings as an older person. He wanted to know if I thought the elderly were treated fairly. Of course the elderly hadn't really existed as a group before 1935. Most of them were part of the family. Their children grew up and the children were living with them until they died. Those that were poor and alone would go to the poorhouse. I'd tell him how, judging from the way old people were taken care of years ago, they are much better off today. If older people were given sufficient funds, they could look after themselves. I think a man is entitled to a pension or Social Security when he reaches the age that they set. From then on he should draw his pension and be allowed

152

to make whatever he can if he wants to make money. If they had the money, a good many could take care of themselves, at least up until eighty-five or ninety, until they broke a hip or something. Of course, women can take care of themselves better than men because women know how to cook and keep house whereas men don't. I sometimes wonder whether the younger generation realizes that the old have much more experience and probably a better "education" then they have, at least in certain things.

The average person hates to grow old. It's one reason why today women spend so much time in the beauty parlors trying to ward off the effects of old age. Men, too, hate to admit that some of the things they used to do fifteen or twenty years ago, they can no longer do. Sure, I wish I was younger. I'd like to be in my forties, but you know it can't happen. I think as you get older the years seem to go by faster. When you're young you can't get old fast enough. You want to be sixteen so that you can drive a car, and then you want to be twenty-one. But when you get older the time flies. Of course, when you're seventy-nine or eighty, you're not going to live too much longer. I think you worry when you're in your sixties whether you'll have enough money when you retire. When you get up around eighty, you realize that you only have a few more years to live, so you don't have to think of saving for your old age. My father was thrifty, my mother was thrifty, and I was never a spendthrift, but I never was tight. Now I find that I probably am less thrifty because of my age. I figure, you know, you haven't too many more years—three, four, or five. You can enjoy yourself.

I don't have any regrets. I'm satisfied with my life. I've helped people. When I was teaching we had a school paper called 'Janus,' and somebody on the paper interviewed some prominent graduates of the high school. They interviewed this fellow, his name is Walton, Steven Walton, the owner of the Lanchester Herald and Echo. And he said in the article, "The teacher I remember best of all was Mr. Foster. He was a great kidder. From him I learned poise and confidence." And of course, I knew what he meant. If you came into my class and you were a bashful kid, having been bashful myself you would have been my pal. I'd get you out of your shyness because I'd never embarrass you. If I'd ask you a question and you didn't know the answer, I'd get you out of it quickly. A good many of these kids, if you handle them gently, they come around.

I have always been tactful, very tactful. When I got my recommendation from my history teacher at college, he mentioned that—that was the gist of the recommendation. When I was teaching at Collaton

High, I was often the peacemaker. I recall one incident. The principal asked me to come down to the office. He asked me if I knew any of the Jewish war veterans. I said, "Yes, a few of them." So he told me what happened. This Irish Catholic teacher had let his anti-Jewish feelings explode. So the principal said, "They're coming tomorrow, and I was wondering if you would come down to the office and be with me when they come." So I said, "Yeah, I'll come." I practically had the whole day off: he sent for me early! All I did was roam around the corridor. They came in around eleven, I guess, and I was in the office with the principal. The principal opened the door and the first fellow says, "Hello, Ted!" The principal said, "You know Mr. Foster?" The follow says, "Sure I know Mr. Foster, everybody knows him." So anyway, the conversation started. They explained what happened. I said:

> Yes, he never should have said what he said. But one thing I
> know, he's learned his lesson. He'll never say it again, or any-
> thing like it.

They said to me, "Well, that's all we wanted to know, Ted." And so they went out feeling happy. The principal was happy, everybody was happy. They always used to call on me.

Yes, I was diplomatic. I remember I was down on Cornwell Street and I was in Turcotte's Diner one night with a friend of mine. He lived down that section; he knew the section very well. I walked up to the counter, and when I did, a fellow staggered over to me. And I saw my friend get up from the table. He didn't move. He just got up. So the fellow, the drunk, looks at me and he says, "I'm from South Boston, I am!" And when I said, "I'm from East Lanchester!" the guy started to laugh. He threw his arms round my shoulder. My friend sat down. When I got back to the table, he said, "What was that all about?" So I told him what happened. Just saying the right thing. The fellow probably didn't know that East Lanchester had a tough reputation, too, but it struck him as funny.

I think as you get older, you're wiser. Anybody who's close to eighty and doesn't know himself fairly well, there's something wrong with him. I'm flattered that people find me interesting. I remember a couple of girls I met up at Hampton Beach. We sat and talked. When she was leaving, one of them said, "You have a very interesting person-ality." And I've heard it so many times that it has helped me through-out life. I think Graham finds me interesting.

154

Oh well. Enough of this. That's his car turning into the driveway. I figure we'll probably have a pleasant afternoon, talking. One thing I enjoy is good conversation.[1]

---

[1]Edward died two weeks after this meeting.

CHAPTER VII

TOWARD A THEORETICAL PERSPECTIVE ON THE GEOGRAPHICAL EXPERIENCE OF OLDER PEOPLE

It is fun to discover lawfulness, and a neat set of experiments that solve a problem can and does produce peak experiences but puzzling, guessing, and making fantastic and playful surmises is also part of the scientific game and part of the fun of the chase.

Abraham Maslow

FIVE PEOPLE, FIVE PLACES, ONE LOCATION

The opening account of Stan's experience and the preceding vignettes have revealed the diversity of the participants. All resided within Winchester Street, but each had established a unique lifestyle and relationship to the area.

Stan was superficially morose and yet, within himself, quietly defiant. His life had been oriented in terms of his work role. When he retired, he substituted the routine of his daily rounds of the bars for the regularity of employment. As he grew older and more sickly, the environment became progressively more constraining. Sadly, and with a trace of bitterness, he damned fate for his predicament, but he adjusted. Resolutely he clung to life despite a weakening body and crippling loss of role which socialization had conditioned him to accept as inevitable.

For Stan, Winchester Street was primarily a functional arena, a familiar physical setting defined as much by his walking range as by any special social affinity. The important places were the bars, havens in a sea of traversable space. The bars were not only an important functional resource but also the focus of a barroom society with which he identified. During his hunting days he had spent much time and made most of his friends outside the area. As he became less mobile his involvement within the neighborhood necessarily increased. Winchester Street became the setting which circumstance decreed as the site of his confinement.

Marie's lifestyle was focused on her family and deeply committed to a peer group social network within the French community. She was more overtly defiant than Stan. Active, outgoing, and doggedly optimistic, hers was not a dull acceptance of constriction imposed by advancing years. Aggressive denial of changing personal capability, self-righteous condemnation of the world, and retreat into defensive reverie characterized her response. Contemplating the richness of her past provided a sense of identity and a source of sustenance.

Winchester Street was, for Marie, almost sacred space. It was the home of the French community, and she was very much at the heart of this community. Her home, the church, and the homes of her friends were locations of special significance set within a total context with which she felt an intimate bond. Winchester Street was both a depressing and supportive place. It was not the neighborhood she had known. The demise of the Imperial Street stores and the social invasion of the "big building" symbolized an oppressive chaos in the space. But at the same time, physical reminders such as the memorial to her son, in conjunction with the social network she maintained, preserved the space of a more auspicious past. This Winchester Street could never be destroyed. It would always be the focus of her incorporation of the locale within her lifespace.

An easygoing jovial flamboyance permeated Raymond's lifestyle. Mindful of the danger of overextending himself, he would rest more, but he limited the frequency rather than the range of his activities. He had established a comfortable equilibrium within his world. He achieved fulfillment from a perceived identity within the local community, participation in the worlds of his family, and deeply rooted fatalism stemming from faith in the beneficence of God. He lived "day for day." Personal ebullience laced with a tinge of egotism belittled the impact of time's passing.

Like Marie, Raymond identified strongly with Winchester Street as a social and historical context. He was part of a community which had occupied and "possessed" this space. The church, the neighborhood stores, and the school were important symbols of this occupancy: physical deterioration, the encroachment of the lumber companies, and the influx of alien populations signaled the loss of this possession. However, Raymond's Winchester Street was essentially the people: his affinity for the space was an affinity for a social order which had existed, rather than a physical setting. Winchester Street was a social space.

Evelyn possessed a less exuberant sense of equanimity. Her family had always been the focus of life. As she grew older she retained her role as matriarch in an ever expanding family network. She was open, trusting, and friendly, but distinctly pragmatic. There was no ranting at the vicissitudes of time, merely calm acceptance of increasing constraints within her life. There was a gentle rueful sadness at changes in the world order, but the future was not to be feared and the past was a reservoir of experience to be savored.

Winchester Street provided a familiar intimate context in which she had carved her own unimposing niche. The space was known in terms of the paths she had traced during her walks, the stores she had patronized, and events she had witnessed which had generated a subtly and finely differentiated landscape within her awareness. For Evelyn, this context was also the locus of the French community. But beyond this, it was where her husband had lived and died, where she had raised her family, and where her friends of many years resided. The social history of her involvement was part of her identification with the space. There was a sense of belonging stemming from an implicit bond with people she rarely saw but whose very presence imbued the space with a supportive aura which made it truly "home."

Finally, Edward, a loner, was a placid thoughtful reflective man, whose horizons extended far beyond the locality. Education projected him into a different social world. Blindness and failing health resulted in a shrinking physical lifespace but he maintained a quiet resolve; adjusting to the loss of his treasured car, continuing his walks in the cemetery, and accommodating to changing capabilities. He accepted rather than fought his aging. Possessing a proud self-knowledge, he derived solace and reaffirmation in contemplation of his life.

Edward inhabited an entirely different Winchester Street from the remainder of the participants. As a physical setting the neighborhood was known with respect to the opportunities it provided for movement: the gaps in the fences and the traffic lights were crucial features of the space. The cemetery was a particularly important domain—space in which he could roam at will. As a social context, Edward's Winchester Street was not the home of the French but the preserve of an English community which had flourished during his youth. Here he had played as a child and forged his first friendships. So it was space laden with historical significance—the stimulus for myriad fond recollections.

159

During his middle and later years Edward identified with a social world beyond the confines of the area. Consequently, even though he resided in the neighborhood for most of his life, he became in a sense, an outsider.

Clearly, each of the participants has a unique personality and has developed a lifestyle consonant with his or her personal adjustment to the aging process. Each person relates to the Winchester Street setting in a highly individualistic manner. Shared space is used differently and has different meaning for each older person. Moreover, there is great variation in the degree to which the participants are restricted within the Winchester Street environment: all are, to some degree, involved in geographical contexts far beyond the confines of the neighborhood, both in space and time. In every case geographical experience is revealed as an extremely complex melange of environmental involvements. This leads us inevitably to the central question posed by my study: is there any commonality in geographical experience? Close scrutiny of the participants' lives suggests it is possible to make some structural generalizations about the manner in which these older people inhabit the spaces and places of their lives. It is useful to distinguish four modalities of geographical experience: *action, orientation, feeling,* and *fantasy.* We turn first to consideration of action.

ACTION

Action, defined as physical movement in space, is the most easily observable modality of geographical experience. It is the outcome of interaction between human potential and environmental facility. Actions are a physical expression of the individual's attempts to satisfy needs and aspirations which require a change of location. Action may be considered on three levels, distinguishable both in terms of geographical scale and frequency of occurrence: immediate movement, everyday activity, and occasional trips.

Immediate movement is a function of bodily agility. It involves the ability to move around within the proximal physical setting. The participants possessed diverse capabilities in this regard. Even before he entered the hospital, Stan was hunched and wasted, his legs stiffened and he shuffled ponderously: merely climbing on a bar stool was a laborious and uncomfortable process. By contrast, Marie was agile and quick; her fingers remained supple. She could reach for a cupboard or crouch to hem a skirt without discomfort.

160

As physiological and health constraints upon agility increase, environmental barriers become more significant. Stan came to dread the ice, could not handle steps on buses, had difficulty negotiating stairs, and hated escalators. Edward's blindness accentuated the barrier effect of the step at Aldea's Restaurant, and Evelyn's admission that she "...wouldn't mind walking if the weather and the sidewalks were good," revealed a similar limitation.

Acute awareness of changing potential seemed to strengthen resolve to delay crippling loss of mobility. Stan would constantly exhort himself to action. "You got to keep moving," he would explain. Evelyn, too, subscribed to the desirability of activity: "...they tell you, many of them tell you, the more you keep going the better it is for you. I think that's true." And the same philosophy underlay Raymond's rueful commentary on an ailing friend, "He's slipping, he's slipping fast because he don't get out."

Adjustments were made to compensate for unavoidable limitations. Stan would seek directions in the supermarket to save unnecessary walking. He would urge me to seek parking spaces as close as possible to our destination, and encourage me to park near to the sidewalk in order to minimize the possibility of tripping on the curb as he got out of the car. Edward had adjusted to his reduced capabilities while out walking. "...I'm more careful where I put my feet down," he explained. There was also a judicious avoidance of unfamiliar physical environments which might present physical barriers. As Edward phrased it, "You tend to stick to the same routes."

Overall, the participants reveal a cautious pragmatism adjusted to individual circumstances. Evelyn liked to remain active, "I always find something to do." At the same time she was aware of her reduced capability: "I was supposed to have them here for a New Year's party and get-together, as the mother. And I couldn't, see, I get tired easily." Raymond, also, mindful of his weak heart, acknowledged a need to "rest up a little bit" after strenuous activity. "I ain't no spring chicken any more, and I have to stop and think now." But perhaps Stan's words best convey the sense of delicate balance between activity and rest which characterizes the participants:

If I sit still too long my legs get stiff now; you know, they get all numb and that...cramped up. When I walk too much they get too tired.

Changes in individual capability are reflected on a larger scale in patterns of *everyday activity,* defined as routine service, social, and recreational trips. Progressive spatial withdrawal was a dominant theme here. When I first knew Stan, he was visiting fewer bars and restricting himself to those a short distance from his home. He no longer went hunting and more time was spent in his apartment. Marie grudgingly admitted, "I don't walk no more because I can't go down street, that's too far, but I walk." Instead of strolling "down city" she spent more time with friends residing within a few doors. Edward was restricted by his blindness and the loss of his car. In some respects his previous high level of mobility served to accentuate the severity of his restriction, even though he was still able to wander around the neighborhood. Evelyn also acknowledged withdrawal. "Well, I go out less." "I used to walk down city. Now I don't think I'd attempt it." And she made fewer social visits, "I don't go out to see many people around here. No, they come in to see me."

Reduced individual capability is accentuated by a pervasive fear to venture abroad, especially at night. "I would never go out at night," admitted Edward. Marie echoed his sentiments: "I'm afraid to go out at night. I don't go out at night no more." Even Raymond exhibited a caution born of insecurity: "When it becomes no more safe to be on the street, to walk the streets or anything else, then you don't dare go out." And Evelyn was fearfully adamant: "I'm never too afraid as long as I'm closed in and the doors are locked." Stan provided the only exception. He had no apparent reservations about wandering from bar to bar in the evening.

Each participant attempted to minimize or compensate for a more limited everyday activity orbit: none was totally quiescent. Several strategies were involved in their response. One was *substitution*. Increasingly, use was made of relatives and friends to secure rides to otherwise inaccessible locations. Evelyn's sister, Lucy, would take her shopping or transport her to church; Marie's friendship network would furnish rides to "card parties" and other social functions; Edward would solicit rides from relatives; and Stan was adept at scrounging rides from acquaintances on the bar circuit. Relatives, friends, and acquaintances were also utilized to make "surrogate" trips. Examples include Evelyn's use of her son to convey hot soup across the street, Marie's use of her granddaughter to undertake her shopping, and Stan's efficient use of "Steve" and "Rick" (the barmen) and other friends to run errands. Third, the use of the telephone provided an important substitute for face to face contact. Evelyn's lengthy conversations with Mrs. Duvalle came to take the place of a social trip to the house down the street.

162

A second strategy was *routinization*. The lessons learned through repeatedly traversing familiar space facilitated continued participation in environments which might otherwise have been almost impossible to negotiate. Edward's awareness of the lay of the land in the cemetery, his knowledge of the neighborhood, and his intimate familiarity with downtown Lanchester, enabled him to negotiate these environments in spite of his blindness. Stan's travels from bar to bar were facilitated by intimate knowledge of habitual paths. Routinization seemed to provide an ingrained sense of distancing, awareness of the appropriate places to stop and rest, a reliable basis for judging his own potential.[1] When confronted with an alien environment—in Stan's case, the New Lanchester Center—this potential could no longer be utilized.

Finally, on many occasions, there was *defiance*—a vigorous refusal to be restricted. Edward would not be denied his walks in the cemetery even though this meant negotiating a busy street and squeezing through a narrow gap in a fence. Marie doggedly insisted on maintaining a taxing schedule of social commitments. And several times when we traveled to his old haunts it was obvious that Stan was extending himself to the limits of his endurance in order to cover as much ground as possible.

Concern with everyday activities sometimes blinds us to the significance of the unusual in people's lives. Four of the participants took *occasional trips* over extremely great distances. Indeed, it appeared that the propensity for such trips had increased as they had grown older. Often such excursions were at the behest of children who had relocated far from Winchester Street. Marie visited her daughter in Florida and traveled to vacation in Gary, Indiana. Raymond spent several months on a trip to Arkansas and North Dakota, and Evelyn traveled to her son's home in Arizona—"My first time I ever flown, too!" Occasional long distance trips were not limited to family visits. Edward achieved spatial liberation through summer vacations at Hampton Beach, and both Evelyn and Marie joined age peer group friends on vacation excursions which often took them far from home. Occasional trips were

[1] As Tuan has expressed this:

"In carrying out. . .daily routines we go regularly from one point to another, following established paths, so that in time a web of nodes and their links is imprinted in our perceptual systems and affects our bodily expectations. A "habit field," not necessarily one that we can picture, is thus established: in it we move comfortably with the minimal challenge of choice."

Yi Fu Tuan, "Space and Place: Humanistic Perspective," in Christopher Board, Richard J. Chorley, Peter Haggett, and David R. Stoddart, eds., *Progress in Geography: International Reviews of Current Research,* Vol. 6, London: Edward Arnold, 1975, p. 242.

anticipated with excitement and fondly remembered. Only Stan did not enjoy the liberation of travel far beyond the confines of Lanchester.

In sum, it emerges that actions are undertaken at a variety of scales ranging from movement within the individual's bodily range, through routine activities within an everyday activity orbit, to journeys of many hundreds of miles (Figure VII.1). In addition, actions vary in frequency. On the level of immediate movement action is almost continuous. The routine trips of everyday activity are less frequent, and the occasional trip is, by definition a comparatively rare occurrence. As the participant's experience reveals, restriction on one level is not necessarily accompanied by a curtailment of action on another. It is entirely consistent for Edward to be restricted on the immediate and everyday level and yet free to travel to Hampton Beach on the larger scale. Different constraints operate at the three levels. On the level of immediate movement physiological capability is crucial. On the everyday scale such limitations are supplemented by environmental barriers and social constraints. Yet on the level of occasional trips, once the space to the transportation terminal is traversed, economic circumstances may provide the only constraint.

ORIENTATION

Actions are rarely context free. Both immediate movements, such as moving around the home, and patterns of activity on a larger scale— the everyday trip to the store, even the occasional transcontinental vacation—are premised upon awareness of the form of the world. This awareness provides a frame of reference, orienting the individual within a geographical lifespace. *Orientation* as a modality of geographical experience involves the differentiation of space through the use of schemata—mental images or representations of the known environment. Schemata provide a cognitive structuring or coding of the individual's milieu, a mental template which sets the limits and defines the possibilities for action.[2]

There is a structural similarity among the five participants in the way they orient themselves within their geographical lifespace. Each functions in relation to a *personal schema* which provides a basic

[2]There is clearly much overlap between the notion of schemata and a rapidly growing literature in the burgeoning field of environmental psychology concerned with "mental maps," "images," "cognitive representations," and a host of similar terms. See, for example, Gary T. Moore and Reginald G. Golledge, eds., *Environmental Knowing*, Stroudsburg, Pa: Dowden, Hutchinson and Ross, 1976.

Immediate Action

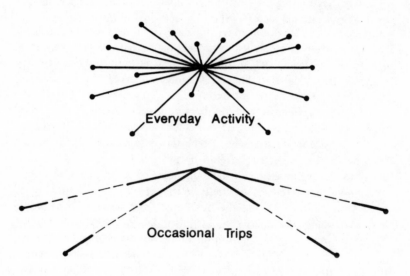

Everyday Activity

Occasional Trips

Figure VII.1 Three Levels of Action

psychobiological orientation. A personal schema facilitates all levels of action. It structures the individual's "lived space" in terms of:

> both a distinct coordinating zero point which depends upon the place of the living man in space, and a distinct axis system which is connected with the human body.[3]

This schema enables the individual to maintain a sense of balance, to distinguish left from right, back from front, horizontal from vertical, and near from far. The existence of a personal schema is fundamental to the individual's being in his world, but as the research process did not provide information with respect to this domain it will not be further elaborated.[4]

Of more concern in this context are an array of *specific schemata* which the participants were found to utilize. These were purpose specific cognitive representations of the environment in terms of paths or routes. Negotiating the environment involved the use of specific schemata as navigational aids. To Stan the local environment was a series of resource nodes (bars) embedded in space to be negotiated. Traversing this space involved a detailed awareness of the path to be taken under diverse circumstances. When there was snow he kept to the right of the path, as he trudged to Steve's, in order to avoid the icy patches. Around lunchtime he would be especially conscious of the traffic on Easthill Street. When it was hot he would take a shady route. At all times he would be mindful of the cracked paving outside his home. Edward also employed finely differentiated schemata of this nature. Walking in the cemetery I found he possessed an uncanny sense of where to tread. Knowledge of the sequencing of traffic lights, awareness of the holes in the fences, sensitivity to the areas likely to be muddy, all guided his actions. Taking the bus home from downtown he would invariably alight on Imperial Street rather than at the "Four Corners." The stops were equidistant from his home but, as he explained, if you came via the "Four Corners" you had to walk slightly uphill, often in the face of a stiff prevailing wind which would gust along Durham Street.

---

[3]Otto F. Bollnow, "Lived Space," in Nathaniel Lawrence and Daniel O'Connor, *Readings in Existential Phenomenology*, Englewood Cliffs, N.J.: Prentice Hall, 1967, p. 179.

[4]For elaboration on basic features of human spatial orientation the reader is referred to Ian P. Howard and W. B. Templeton, *Human Spatial Orientation*, New York: Wiley, 1966.

Specific schema were not limited to ambulatory activity. Edward's caution to me as we drove up Treeside Street, enabling me to anticipate the potholes, illustrates how much schemata are utilized in all forms of environmental negotiation. The import of such schemata in facilitating orientation is even more poignantly revealed by another rueful observation he made:

When I'm out in a car, the people I'm with, they all know I know the roads, but the minute we hit a new highway I'm lost. If it's an old highway I can tell them where to go but on the new ones I can't.

While the participants employed an array of specific schemata in traversing the environment, it was apparent that these were situation specific operational devices superimposed upon a more fundamental differentiation of the environmental setting. Actions were also framed in relation to a general schema, centered upon the home, and differentiated into a series of experientially distinctive domains (Figure VII.2).

Home was the fulcrum. Sharply demarcated from outside, it provided private secure, even sacred space, a realm of total jurisdiction.[5] Life was oriented with respect to comings and goings from this domain. For Marie and Evelyn especially, home was experienced as possessed familial space—a territorial preserve. It was, in Bollnow's terms, "an inviolable area of peace, and thus sharply differentiated from the outside world without peace."[6] For the men, home seemed more functionally defined. Spartan furnishings suggested a base of operations, a place to sleep rather than a fully personalized domain. For Edward especially, the Winchester Street home seemed to become merely a location to hang his hat.

The *surveillance zone* formed a second distinctive realm within the participant's general schema. This zone was defined by the field of vision from the home. This was the zone Stan would wistfully monitor from his window on an icy day. Marie felt threatened by the intruders who moved into the "big building" but was far less concerned about similar encroachments farther down the street. Raymond became upset, not when the lumber companies started moving into the neighborhood,

[5]The idea of home as a distinctive inviolable zone of space has been explored by a plethora of writers. See, for example, Gaston Bachelard, *The Poetics of Space,* Boston: Beacon Press, 1969, pp. 38-73; Mircea Eliade, *The Sacred and the Profane,* New York: Harcourt, Brace & World, Inc., 1959, pp. 56-58.

[6]Otto F. Bollnow, *op. cit.,* 1967, p. 182.

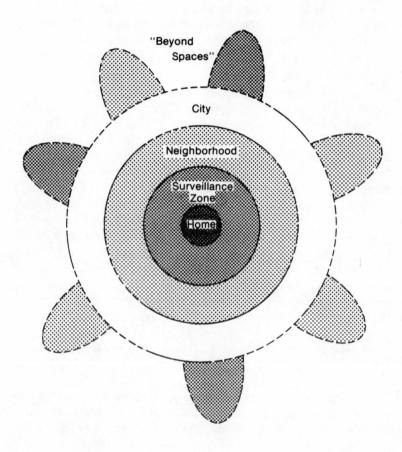

Figure VII.2 The General Schema

168

but when they actually transgressed within the zone he could view from his rear porch. In this domain partial personal jurisdiction was assumed. Unlike the home, this was not inviolable space, but was characterized by a watchful reciprocity among neighbors. Here one developed mutually dependent relationships. It was natural for Evelyn to provide soup for Mrs. Royale across the street and to monitor the children passing by. For Raymond, this was the zone in which he would keep an eye on Mrs. Boudreau across the street, and it was only to be expected that his neighbor would watch his window to check that all was well with him. The absence of close neighbors tended to reduce the significance of this zone as a social context for Stan. The surveillance zone also seemed less important in Edward's general schema. In part this was a result of his blindness, but it also reflected his social orientation within a wider geographical context.

The surveillance zone was embedded within a third zone of the general schema, the *neighborhood*. Neighborhood space did not seem to have clearly defined physical limits even though, when pressed, participants were able to assign crude spatial boundaries. [7] Neighborhood seemed more functionally than formally defined; a fuzzy synthesis of spatial designation and social meaning. For the French participants the neighborhood was distinguished as the catchment area of St. Mary's Church and the preserve of the French community. Involvement was neither as intensive nor as personal as within the surveillance zone, but there was a sense of social affinity and identification with a local community. Neighborhood transition was perceived as a threat to the social identity of this space.

Neighborhood space had a somewhat different constitution in Stan's and Edward's general schema. For both, this realm was more strongly defined in terms of walking range, as a physical setting rather than a social entity. St. Mary's Church, so important to the French, had no significance as a social focus. Socially, Stan's neighborhood was framed in relation to a "barroom society": Edward's was differentiated in terms of childhood memories of an English community.

---

[7] During the field research an attempt was made to have the participants delimit the boundaries of their "neighborhood." After some hesitation, Evelyn and Edward were able to accomplish this, and the remaining participants were able to designate partial boundaries. However, it was clear that they were not used to considering the boundaries of the neighborhood in such specific spatial terms. Further probing revealed a series of experiential boundaries: responses varied according to "what *I* defined" as neighborhood. The attempt to designate clear spatial boundaries for this zone was thus abandoned.

Moving progressively farther from the participants' homes, the general schema changed from a series of continuous and integrated zones to a more disjointed and fragmented differentiation of space. The *city* lay beyond the neighborhood. Both Marie and Evelyn would refer to trips "down city" thus clearly indicating its conceptual separation from the local context. This "Lanchester" beyond the neighborhood was known in several ways; as an array of intimately experienced environmental fragments, as a spatial entity, and as a distinctive social milieu. Awareness of the city as a complex of intimately known settings was closely linked with personal history. For Stan such fragments included localities he had first lived in on his arrival in Lanchester, the area around the mill where he worked, the bars he patronized, the stores on Cornwell Street, and the Pinewood Health Spa. Edward's schema included the school where he taught, the downtown stores he patronized, and the Diner's Haven. Indeed, for each of the participants, the city was embraced within the general schema as a veritable collage of experienced haunts. Intimate knowledge of individual settings was set within an overall sense of the city as a spatial entity. When we drove neither Stan nor Edward had any difficulty directing me around Lanchester. All the participants could clearly describe the location of various stores and provide detailed information on the shortest route from place to place, thus suggesting a comprehension of the city as a spatial entity. Finally, awareness of the city was permeated by an ingrained sense of affectionate identification with the old Lanchester. The role of places which had been important in the past was accentuated in the inclusion of this realm within the general schema. The New Lanchester Center did not seem to be part of the essential Lanchester: consequently it was not prominent within the city domain of the participants' general schema.

The general schema embraces spaces far beyond the confines of the city; a host of what may be termed *"beyond spaces"* are incorporated. Stan's general schema included his former hunting haunts; New Britain, Connecticut; New Hampshire; and vestigial awareness of Poland. Marie's embraced Florida, Arlington, Gary, and Quebec. Evelyn's included spaces ranging from Orion, a local suburb, to the Arizona township in which her son resided. Raymond's "beyond spaces" included Arkansas and North Dakota, and during our association Japan came to be incorporated. Indeed, all the locations of the participants' lives were integrated within their general schema. As is apparent from the vignettes, some of these "beyond spaces" were especially significant locations in the participants' lives—places with which they experienced a high degree of personal affinity.

Thus far, in the desire for clarity, a simplified characterization of orientation has been presented (Figure VII.3). At this point it is useful to briefly note some of the complexities of this modality of geographical experience.

First, orientation involves both *conscious* and *implicit* awareness. Over time, conscious awareness may, through a process of internalization, become implicit. As Griffin observed:

> We are likely to know without conscious effort or counting such topographical relationships as three steps from the bed and it is time to reach for the knob on the bedroom door, six more steps to the left brings us to the stairs, after four steps down we reach a landing and must turn right before continuing down ten more steps to the downstairs hall.[8]

As he walked to Steve's Bar, Stan did not have to register consciously every obstacle, every feature of the space he was traversing. He did not have to record consciously every crack on the sidewalk in order to avoid tripping. A kind of "automatic pilot" took over and guided his movements. The existence of such implicit awareness in part explains why the participants experienced little difficulty in negotiating familiar settings.

The issue of *scale,* as reflected in the level of detail of schemata, further complicates the notion of orientation. As Tuan has noted:

> Experience constructs place at different scales. The fireplace and the home are both places. Neighborhood, town and city are places; a distinctive region is a place, and so is a nation. . . They are all centers of meaning to individuals and to groups. As centers of meaning the number of places in the world is enormous and cannot be contained in the largest gazetteer.[9]

The participants possessed a hierarchical awareness of place: the same location was frequently incorporated within individuals' schemata on a continuum of scales. Edward's downtown was known on one level in terms of detailed micro level schemata highlighting uneven paving, high

[8]Donald R. Griffin, "Topographical Orientation," in Roger M. Downs and David Stea, eds., *Image and Environment: Cognitive Mapping and Spatial Behavior,* Chicago: Aldine, 1973, p. 299.

[9]Yi Fu Tuan, "Place: An Experiential Perspective," *Geographical Review,* LXV:2, 1975, p. 153.

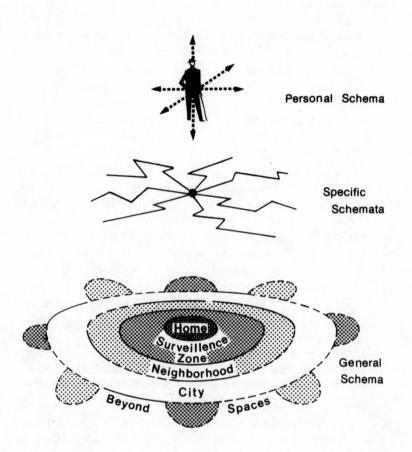

Personal Schema

Specific
Schemata

Home
Surveillence
Zone
Neighborhood
City
Beyond   Spaces

General
Schema

Figure VII.3 Three Dimensions of Orientation

172

curbs, and the traffic lights; on a larger scale in terms of the spatial arrangement of stores; and as a total entity, as a region within his schema of Lanchester as a whole.

Familiarity seemed to facilitate the maintenance of detailed subtly differentiated schemata with regard to frequently traveled routes and zones adjacent to the older person's home. However, there was no simple linear decline in the detail of schemata with increasing distance from this point. Edward had an intimate awareness of the form of the environment adjacent to his home, but he also possessed detailed schemata of routes in the area around the farm he visited. Raymond possessed finely differentiated schemata of familiar paths and zones in the area around his house. An Arkansas garden was also known in intimate detail. The carrots were just beyond the potatoes next to where the tomatoes were planted. In both examples the settings were many miles apart. The space between these places was in each case known in far less detail.

Third, it is useful to emphasize the *territorial* aspects of orientation. The general schema in particular, orients the individual in relation to spaces distinguishable in terms of degree of jurisdiction. Home, the lair, is a place of total jurisdiction; the surveillance zone is a domain of partial personal jurisdiction; within the neighborhood there is a sense of social affinity; the city is a realm of affectionate identification; "beyond spaces" are generally less intimately possessed, but on occasion evoke strong personal identification. Each zone is prescriptive of certain types of behavior. The participants varied, however, in intensity of identification within the various realms. A model of linear decline in sense of possessiveness with increasing distance from home was appropriate for Marie and Evelyn. But Raymond possessed a stronger sense of territorial imperative in the surveillance zone. Stan's sense of territory seemed weak. Edward, too, did not reveal a sense of possessiveness with respect to physical space of the present. His territory was less spatially bounded. Indeed, his cognitive structuring of different spaces within the city reflected a less place based orientation than the other study participants. His less territorially defined schemata seemed to emanate from his more cosmopolitan lifestyle.

Finally, it is helpful to acknowledge the *diverse processes* underlying the emergence and maintenance of schemata. Here it is useful to employ Downs and Stea's distinction among direct sensory, vicarious, and inferential components.[10] In Stan's schemata, the New Lanchester

---

[10]Roger M. Downs and David Stea, *op. cit.,* 1973, pp. 22-24.

Center was in part constituted from the sensory impressions he received during our visit. It was also influenced by vicarious knowledge acquired from reading the newspaper and viewing the shopping mall on television. In addition, the constitution of this space within his schemata was partially inferential. Experience of large stores and new buildings in other contexts facilitated making inferences about the space (for example, that there would be too many stairs). Even before her son headed west, an inferential awareness of Arizona as a "beyond space" was a component of Evelyn's general schema. It was known as a sunny spacious retirement area. All manner of associated characteristics of the state could be inferred—the scenery, the lifestyle, and the friendliness of the people. When her son migrated she became able and more eager to supplement this inferential awareness with vicariously experienced information. She would watch the weather forecast for the state, she would receive letters and postcards from her son. Finally, a year before I met her, she had been able to enrich her general schema through direct sensory experience when she traveled to Arizona.

FEELING

Feeling constitutes a third modality of geographical experience. Space is imbued with meaning: places possess distinctive auras. Indeed, locations "live" by virtue of emotions they evoke within the individual. Feelings about place may reflect sentiments ranging from dread to elation. Often they are amorphous, multifarious, or inchoate. On an intuitive level it is clear that feelings associated with place are an integral component of the participants' geographical experience. The problem arises when an attempt is made to develop beyond this general statement. The experiential essence of sentiment is lost in the process of communication:

> We instinctively try to solidify our impressions in order to express them in language. Hence we confuse the feeling itself, which is a perpetual state of becoming, with its permanent external object, and especially with the word which expresses this object.[11]

Clearly, it is necessary to acknowledge the ultimate privacy of feelings. However, intensive and prolonged contact with the participants facili-

[11]Henri Bergson, cited in Yi Fu Tuan, *op. cit.,* 1975, p. 218.

174

tated a degree of empathy which made it possible to derive some sense of their feelings about the places of their lives.[12]

Feelings about places could be classified in terms of their degree of permanence. Some were *immediate,* highly situation specific, and relevant for only a short duration. Illustrations here include Stan's initial reaction, his sense of confusion and alienation, upon entering the New Lanchester Center, and Edward's feeling of somber melancholy when he talked about the railway accidents as we stood overlooking the tracks. For a brief period the space was imbued with an aura of tragedy. Other feelings, although *temporary,* have somewhat longer duration and, in many instances, a repetitive consistency. For Stan, the neighborhood at lunchtime was invariably sensed as bustling and crowded. At night the same place was quiet, even tranquil. When it was icy this same environment became menacing. For Marie and Evelyn, Winchester Street was friendly space by day but frightening at night.

The participants' lives were also enriched by an array of more *permanent* feelings. There was a stability in the deeply ingrained emotional identification underlying the French participants' attachment to the Winchester Street area. Wherever they resided this space would always be home. Such feelings expressed an intimate bond established over the years which imbued the neighborhood with a distinctive persistent aura—a pervasive sense of the "urban village" quality of the neighborhood past. Often this attachment could only be vaguely articulated but it was deeply sensed as was revealed on the many occasions upon which such sentiment was used to justify overriding the logic of more "objective" assessments of the local environment.

There is a second way in which feelings about space may be classified. All feelings are individually experienced, but some are shared, inasmuch as they are consensually generated and reinforced. It is helpful to distinguish between *personal* feelings stemming directly from an individual's unique experience of a physical or social context, and *shared* feelings, involving the mediation of other persons in sustaining an intersubjectively experienced sense of place.

Acknowledging the uniqueness of individual experience is essential in understanding personal feelings about places: biography provides a reservoir of emotional identifications.

[12]Of course I can never know how accurately I was able to capture the essence of these feelings, even though I frequently asked the participants to articulate their emotions and was constantly probing to ascertain whether my interpretations were meaningful to them.

There are, for example, privileged places, qualitatively different from all others—a man's birthplace, or the scenes of his
first love, or certain places in the first foreign city he visited in
youth. Even for the most frankly nonreligious man, all these
places still retain an exceptional, a unique quality; they are the
"holy places" of his private universe, as if it were in such spots
that he had received the revelation of a reality *other* than that
in which he participates through his ordinary daily life.[13]

Marie's home, a cosmos in a chaotic world, harbored an array of
feelings. As memorial of birth and death it evoked a sense of quiet
pathos; as the seat of the family, warm, joyful emotions. Other distinctive feelings stemmed from its constitution as workplace and symbol of
achievement. Such feelings were not limited to home. The signpost on
the street bearing testimony to the wartime sacrifice of her son evoked
a sense of loss and imbued the spot with special significance. For Stan,
each of the bars was pervaded by its own distinctive atmosphere.
Steve's Bar, Murphy's, the Half Moon, and Gervei's were homely and
welcoming, Selena's was too "plush and ritzy," and in Laporte's and
Moreno's he experienced feelings of alienation, oppressive crowding,
and a sense of raucous nausea. Indeed, each of the participants imbued
specific locations in their lives with feelings which personalized them.

Personal feelings were also attributed to areas. Stan avoided certain
districts: "The only place I wouldn't go is on High Street there, and
places there they've got all the queers going." Raymond felt a sense of
importance and benevolent possession in the area embracing his former
bakery route:

> So I got to know everybody, every damn one in this neighbor-
> hood, every house, every family. See, they all called me, and
> they still do, a lot of them, "Ray the baker." I used to take
> pastry and cut it up in small pieces, in them days people were
> hard hit, see. And the kids would be playing in the street, and
> I'd cut them all up and spread them out on a piece of paper on
> the tailgate of the truck. Then I'd whistle and say, "Come and
> get it!" Oh, if you don't think they'd come. I was the greatest
> man on earth.

For Edward the cemetery evoked strong personal feelings. In this space
he felt secure. Away from the traffic, this was "free" space—he could
roam at will. At times the cemetery was melancholy; the final resting

[13]Mircea Eliade, *op. cit.*, 1959, p. 24.

place of many friends, it could instill an aura of somber affinity with time past. Most of all, the cemetery was an area which aroused feelings of identification with an adventurous childhood. As he would frequently remark, it had been "like a playground for us." Evelyn provides a final illustration in her discussion of the prospect of leaving Winchester Street:

> I'd love to go to Arizona if the other two were with me. I'd be with my youngest boy. But then I'd have two left back here. And then, I'll tell you why. My husband has done things here. It's awful to talk about; we all expect to die anyway, sometime. We have a cemetery lot. So if something happened way down there, they'd have to move me way back here.

Involvement in a common context, shared value systems, and high levels of interaction, facilitated the development of consensually held feelings about places.[14] There was a commonality in the French participants' feelings about St. Mary's Church. Beyond its mutually accepted aura as a sacred place, it was imbued with a protective identity as a womb for the French community. Evelyn's comments express this more secular sense of shared commitment:

> ...that's a place to go until we're not able to move, or sick or something like that. I give a dollar a week. I think *we* owe it to our church.

Raymond revealed a third facet of shared feelings, a sense of social identification, defining the church as a distinctive social place:

> ...it's in your life. Instead of going to the night clubs, and sit down and watch somebody or other, you're having a good time with your own crowd.

A similar commonality characterized feelings about Imperial Street. A shared sense of rueful loss was accentuated by memories of the physical setting that had been. As Evelyn phrased it:

> This has changed ever since my husband died. The stores were beautiful around here. We had a nice pharmacy here, Lebouf's drugstore, and that was lovely. But see, well now it's Selena's, they have that nightclub.

---

[14]The absence of detailed reference to Stan and Edward in this discussion of shared feelings is intentional. Both participants were very much "loners." Neither strongly identified with a social group which was a source of common feelings about place or space.

A social dimension within this pervasive feeling was articulated by Marie:

> Imperial Street used to be lively. There was everything. There was a dance hall in the corner there, right at the bridge. The club there, we used to go down, dance the quadrille. All right at the back of the drugstore. Now it's dead, there's nothing there.

Although the depressing aura was diversely expressed, shared feelings about Imperial Street evinced a sense of the place as symbolic of the tragedy of neighborhood demise.

There was a historical social bond among the elderly members of the local community, best characterized as a "field of care,"[15] which seemed to be translated into affinity for Winchester Street space. Direct expressions of such covenants between social groups and the environments they inhabit are difficult to detect; indeed, they are only partially self-conscious. The existence of such implicit "fields of care" is suggested by the social/spatial age peer networks which have been identified (Figures III.1, IV.1, V.2, and VI.1). "Fields of care evoke affection,"[16] but this affection may only be fully appreciated when the processes which sustain it appear to be eroding. Marie acknowledged a social identification with Winchester Street, but the sense of community which sustained this field of care was threatened:

> We used to know in the old time when I moved down here, everybody. Everybody know us and we know everybody. They talk to us. Now we don't know. There's no more friendship like there used to be. We used to help each other's family. If one was in trouble, if one was sick, we used to go down and help them, but now there's no—the people ain't friendly.

Raymond, too, was aware of the vulnerability of social attachments to place. He remembered:

> You, your wife, you'd take a walk and go round the block. Now, you come at night time you don't see nobody walking the streets. Oh hell, in nineteen forty everyone used to go walking night time. You used to hang out the fence and go over next door and you talked to nearly everybody.

[15]Yi Fu Tuan, *op. cit.*, 1975, p. 236.

[16]Yi Fu Tuan, *op. cit.*, 1975, p. 236.

He sensed that as social ties weakened the strength of group identification with space was being undermined.

Clearly, there is considerable overlap between orientation and feeling as modalities of geographical experience. Emotional attachments not only provide places with special meaning, they also differentiate them from surrounding space. Conversely, breakpoints in schemata often coincide with boundaries in types of emotional attachment. For example, the border between the surveillance and neighborhood zones of the general schema is delimited not only by the limit of visual acuity but also by a change in the quality of affective identification.

A single setting may evoke a wide spectrum of emotions. Such multiplicity reflects temporal variability in feelings for places. For Marie, home could foster feelings of joy or tenderness, the warm aura of a golden past. Moments later it could be sensed as the site of domestic tragedy. This feeling could in turn give way to an air of bustling industrious activity. Considering a longer time span, for Stan the Boundary Bar and the Rusty Harness were once welcoming. A few weeks before he died they had become alienating rather than friendly bars. To Evelyn the downtown of the past had been intimate, almost cozy space, "I thought it was pretty." By the time I knew her it was sensed as expansive, cold and impersonal.

Feelings about spaces also vary according to situational context. The church was pervaded by one array of feelings when Evelyn attended a funeral but different emotions when she was at a wedding. Similarly, as a place to stroll at night, Winchester Street was, to Marie, dangerous and frightening. As a social arena by day it was a supportive and friendly setting. The same space may foster seemingly contradictory emotions.

FANTASY

In discussing orientation and feeling we have already, on occasion, merged into consideration of vicarious forms of geographical experience. There has been an implicit acknowledgment of geographical fantasy, one of the least probed and yet most fascinating modalities of geographical experience. The participants' lives were enriched by involvement in locales displaced in space and time. The richness of Marie's geographical experience was not so much a function of her hectic pace of contemporary activity, as of her ability to range far and wide in the places of her imagination. As Raymond sat in his kitchen he

179

could project himself into the worlds of his children; he could participate in a North Dakota celebration, reflect upon an Arkansas garden, and even immerse himself in contemplation of life in an Oriental city. Edward captures the essence of such involvement in displaced milieu:

> And sometimes I think about places I went. I'm the type of person that likes to dream a little. Dream, I call it dreaming. I think it's good for you; just takes up your time. I try to keep off sickness and things that bother me, but I'll often just sit and think about the things that happened. I have a tendency to think of what pleases me whereas some people might think of things that make them bitter. I don't have any regrets. I don't look back negatively. Most of my thoughts are on pleasant things. Oh yes. Sure, I often sit and dream.

The term fantasy is not used in any negative, demeaning, or perjurious sense, but as a general designation for a modality of experience which appeared to have particular significance within the totality of the participants' geographical experience. As a prelude to exploring its geographical manifestations it is useful to propose some conceptual distinctions which clarify the nature of fantasy. First, fantasy is a genuine component of a person's experience. Indeed, ". . .the human imagination never invented anything that was not true, in this world or any other."[17] At the same time, there is often only tenuous correspondence between fantasies and the external world. Fantasies are creative "grand fictions" through which the individual molds a separate personal reality. As White observes:

> When I tell the story of my life, it is largely made up of the images I create of the places in my life. Remembering the places and the emotions as I once experienced them is a tricky business. I don't exactly discover the past of my life by remembering events. In fact I invent my past as a grand fiction, the myriad details of which fit into a coherent pattern that is called a self concept. There is much twisting and bending of the original event so that it can fit into the model of what I say I am, and what I say the world is.[18]

[17]Gerard de Nerval, cited in Gaston Bachelard, *op. cit.,* 1969, p. 153.

[18]Ernest A. White, "Environment as Human Experience: An Essay," Unpublished M.A. Thesis, Clark University, 1973.

In this sense fantasy is unreal.

A closely related dichotomy is that fantasy is both contemporary and not contemporary. Vicarious geographical experience exists in the immediacy of the individual's life: it is as much a part of the present as breathing, eating, or taking a walk to the store. However, the content of fantasy need not be contemporary. I can immerse myself in the space of my childhood home, returning to the havens of my youth, or I can project myself into a surrealistic world of the future. Fantasies do not limit me to my contemporary milieu.

From this acknowledgment stems a third, and in this context crucial, feature of fantasy. *Fantasy makes men free.* Poignant poetic illustration of this characteristic is provided by Bachelard:

> A prisoner paints a landscape on the wall of his cell showing a miniature train entering a tunnel. When his jailors come to get him, he asks them "politely to wait a moment, to allow me to verify something in the little train in my picture. As usual they started to laugh, because they considered me to be weak minded. I made myself very tiny, entered into my picture and climbed into the little train, which started moving, then disappeared into the darkness of the tunnel. For a few seconds longer a puff of flaky smoke could be seen coming out of the round hole. Then the smoke blew away, and with the picture, my person." How many times poet painters, in their prisons have broken through walls by way of a tunnel. How many times as they painted their dreams, they have escaped through a crack in the wall! And to get out of prison all means are good ones. If need be, mere absurdity can be a source of freedom.[19]

In fantasy the individual is liberated. Physiological decline, ill health, economic constraints, social alienation, or environmental barriers provide no limitation. The only boundaries are those imposed by autobiography, personality, and the limits of imagination.

In what ways was fantasy manifest as a modality of the participants' geographical experience? Each participant was involved in many different worlds. However, two general types of fantasy could be distinguished. *Reflective fantasy* involved reminiscence, an immersion in environments of the past. This process entailed the reconstitution in consciousness of the places of individual biography through reliving

[19]Gaston Bachelard, quoting a fragment from the writing of Herman Hesse, Gaston Bachelard, *op. cit.*, 1969, p. 150.

events which transpired within them. A second type of geographical fantasy included involvement in environments inhabited by their children or other relatives, and reveries encompassing imagined worlds of the future. This may be characterized as *projective fantasy*. We first consider reflective fantasy.

Reflective fantasy involved the participants in the complete range of environments of their lives. Stan could participate in the Poland of his childhood, New Britain, Connecticut, the many places where he had hunted, and a host of other locales. Marie could be in Quebec, could muse in environments of West Carlton, Chaugon, and other neighboring towns, could travel to Arlington, could transport herself to Florida and Indiana, and could reinhabit the environment that was the Winchester Street of the past. The world was far larger than the contemporary context.

Reflective fantasies were centered on events, and it was this remembrance of event which seemed to evoke remembrance of place. Raymond vividly recalled the neighborhood on the night of the hurricane. Edward remembered Imperial Street and the area beyond in the context of a scary journey home from a "Jekyll and Hyde" performance:

> There was only the mill and a big long house; the rest of it was open. So we'd rush through there. We'd run. . . Boy I was scared! Nothing ever happened. It was just the thought of it.

The railroad crossing was recalled as it was on the Sunday morning of a gruesome tragedy. And Marie, in reliving the time when the little girl next door died of pneumonia, reconstituted her neighbor's house as she perceived it was then.

The manner in which places of the past are constituted in fantasy depends on the way events are incorporated within the person's reminiscences. Reflective geographical fantasies seemed to be more than idle reminiscence. Selective recall of events, and by extension, places, provided support for personal identity. Stan's reflective fantasies took him back to a rural Poland in which images of ploughing with two horses, and strenuous forestry work predominated. This constitution of his childhood milieu was consonant with the work ethos which pervaded his life, "We worked up there. There was no loafing." Marie's memories of her home as a colorful environment in which her children thrived reinforced a self-perception of a loving and successful mother:

And we used to dance. We used to take that table what I have down here, and we put it on the piazza, and we used to dance the quadrille down here in the kitchen. That was the happiness. And we danced all night—sometimes to one o'clock.

Sometimes the places of the past were recalled as a contrast to the present. The hallowed memory of neighborhood past reinforced a sense of belonging and participation in the history of the space. It accentuated disenchantment with the contemporary scene. However, most important, reflective fantasy imbued contemporary place with depth of meaning inasmuch as the stream of its past was incorporated within its experienced identity.

The participants were also able to project themselves into settings spatially and temporally beyond their contemporary environment. Most clearly apparent was a propensity to span vast distances and enter into the environments of their children. Raymond's garden in Arkansas was "tended" with care from a distance of many hundreds of miles. When we talked of this place, often he was not with me:

See, they got like a playground. There's a fence here and then, there's the other fence. Now in between this fence and that fence we make the garden on the right-hand side. And they dig sand and everything else. And there's swings just like the playgrounds. On the other side is where *we—they* keep a horse and cow. And these are all trees. There's twenty-three trees with grass.

And on the day when he excitedly explained the detailed sketch map of his son's Tokyo apartment which he had just received in a letter, he was for a while a participant in Oriental life. For in many ways Raymond was with his children. As he explained, "Your success after you've raised them is *to watch their family grow and take part of it.*" Such vicarious involvement was not limited to Raymond. Marie was often involved in the worlds of her children. Sometimes she "traveled" to Florida. On other occasions she "visited" a Gary home. And Evelyn participated in an Arizona environment:

I watch the map on the TV and they've got cool; and it's going down to about, like in the forties, forty-five. He's in the center of Arizona.

She shivered. In short, the participants' lives were enriched through vicarious participation in the environments of others.

There was also some evidence of a propensity to become involved in more ambiguously construed environments of the perceived future. Several participants would muse on the eventual fate of the neighborhood. Themes of continuing deterioration traced a dismal motif. As Raymond envisioned it:

> The pressure's on. Five years? Ten years? Fifteen years? This neighborhood in another twenty-five, fifty years, won't exist. It'll all be rebuilt. As the neighborhood is now it's marked, it's marked on the plans, down city hall there.

There was an inkling of similar inchoate fantasy in Evelyn's wistful comment, "Maybe there'll be no Winchester Street. Oh, I hope. . ." In addition, on occasion, the participants' geographical fantasy anticipated future visits to unfamiliar landscapes. Even before our visit Stan could anticipate journeying to the New Lanchester Center. Raymond could savor the prospect of a trip to Tokyo. In his musing he could picture the crowded streets, the bustle and profusion of color, oppressive heat and the constant jabbering of an alien tongue.

Clearly, the participants vicariously experience many environments, displaced both in space and time. The possibilities are schematically represented in Figure VII.4. Events in fantasy may be located anywhere within this space/time frame. When Stan recalled his Polish childhood, when Raymond mused on a boyhood paper route, and when Edward was "dreaming" of adventures in the "jungle," all were involved in environments of the distant past. In Stan's case, this environment was also spatially far displaced. For the other participants, childhood fantasies involved geographically proximate settings. When Raymond, Evelyn, and Marie thought of their children, they projected themselves into more contemporary milieu, but in this case the environments were far removed in space. Although there were some indications of vicarious involvement in imagined environments of the future, and such projective fantasy is obviously theoretically possible, the weight of the evidence from the research process strongly indicated that participation in spatially displaced contemporary settings and in places of the past was of much greater significance.

In addition to space/time displacement, personal displacement seemed to be a characteristic of geographical fantasy. As J. B. Priestley once observed:

> It is as though walking down Shaftesbury Avenue as a fairly young man, I was suddenly kidnapped, rushed into a theatre and made to don the gray hair, the wrinkles and other attri-

184

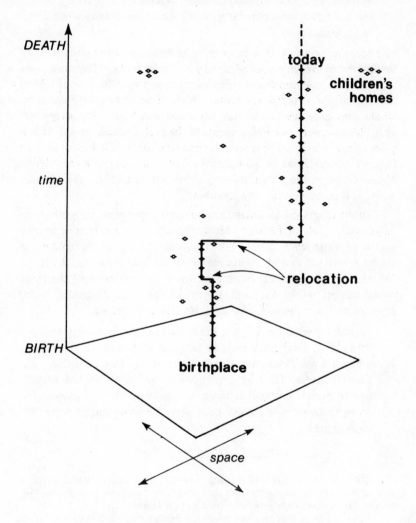

Time/space location of fantasy incidents ⋄

Life path of residence ⌐

Figure VII.4 Fantasy Space

185

butes of age, then wheeled on-stage. Behind the appearance of age I am the same person, with the same thoughts as when I was younger.

The concern here is with a propensity admitted by the participants to involve themselves in places of fantasy as "other selves." The most common process was immersion in an environment as if they were still at the age of the original encounter. When Stan talked of Poland, he would sometimes be lost to me. No longer was he a sickly sixty-nine-year old, morosely describing events of the past. Instead, he was fifteen years of age: immersed in reverie, he was savoring the fresh clear air of a Polish countryside as he ploughed behind two straining horses. When Marie reveled in the joyous dancing in her kitchen of the past she was once again an exuberant young mother.[20]

Often geographical fantasy has no overt expression save perhaps an involuntary smile or furrowed brow, providing in countenance an indication of experience which words cannot convey, as the person becomes immersed in the environments of his life. However, it can be inferred that geographical fantasy incorporates expression of the three modalities previously discussed. Raymond, describing his bakery route, provides a clue to the role of actions. He travels in fantasy:

> . . .and I'd go down Blyton Lane, back of the cemetery, and then down Durham Street, as far as Andrews, as far as the bridge I'd go. Then come down Cutler Street, Tinterne Street, Easthill Street. Then up Cutler Street, and then back down Sentry Street. Easthill to Imperial, up Imperial to Andrews, down to Durham. I'd come back down Andrews and go down Easthill Street.

[20]The words of an elderly hospital patient provide further eloquent expression of both this propensity to take the part of the person of one's past and the conflict between the reality of the present and the fantasy:

"When I walk in that park and see all the children playing their games, I think to myself that I can still do those things. I feel that my body could still be cranked up to play like they play or push some little angel on the swings. The sad part is that I have to keep telling myself that I can't do it. I'll be there studying, watching, thinking I can throw a ball still, I was quite an athlete in my day. Then I tell myself, you're an old man Simpson. You couldn't even stoop over and get the ball if it came your way. Who are you trying to kid? Certainly not the person who handles time. But then I'll be thinking I could just climb back into my twenty-year-old ways."

Thomas J. Cottle and Stephen L. Klineberg, *The Present of Things Future*, New York: Free Press, 1974, pp. 44-45.

The significance of orientation is indicated in his description of his boyhood street:

> In them days there were no houses there, we were the first house on the street. Then there was a stone crusher. . . And then there was a house up by the top of the hill. And then on the other side from the beginning of the street there were houses up as far as us—small houses.

Feelings for place are also involved. Sometimes I talked with Marie about her son's burial in Arlington. As she sat subdued and reflective, seemingly oblivious to my presence, was she experiencing the sanctity of the burial site? Could she sense the pomp and ceremony of the funeral? I suspect she could.

As I came to know the participants it became increasingly apparent that certain contextual *cues* provided important stimuli to geographical fantasy.

Small *artifacts* and mementos provided reminders of places past and stimulated involvement in the environments of children. Evelyn, Raymond, Marie, and even Stan, possessed photographs. Family portraits prominently displayed, often seemed to evoke geographical fantasy. Evelyn would start to talk of Arizona, Marie would journey to Florida. In addition, Marie maintained her bureau full of scrapbooks, a written and pictorial record of her life. When we went to the bureau it was almost impossible to stem the torrent of reminiscences and travels in fantasy which ensued. She would pick up a scrapbook, perhaps one including photographs of a journey to Florida. There was no use protesting. I would have to share the journey. Sometimes she sat alone with her scrapbooks, abandoning herself to solitary reverie. "If I write my life, I'll tell you the truth here, that'll be the best story that you ever can tell." The scrapbooks maintained the integrity of the dream: they provided a reservoir of cues for fantasy.

One afternoon at Evelyn's we had finished tea and were preparing to settle down to watch a favorite quiz show. As we moved to the living room she reached for a small plaque painted with a crude map of the area in Arizona where her son lived. For a few minutes the television was forgotten as she roamed—describing the places she had visited. On his mantle Raymond kept a clock informing him of the time in Tokyo. It was an important cue. "Ah yes," he would muse, "they will be breakfasting now." Postcards from his grandchildren and letters from his children provided a similar stimulus.

The *environmental context* also provided cues to geographical fantasy. Each room, indeed each corner of Marie's home, was a potential stimulus for the resurrection of events and hence place past. I would see a rather somber-looking abode cluttered with furniture and dresses, but this place meant far more to her: it was a storehouse of cues. Lengthy residence also contributed to making home an album of cues for Raymond, for Evelyn, and to a lesser extent, for Edward.

Outside, even though the environment had changed considerably it still acted as a stimulus to recollection. The richness of the setting in this regard was particularly apparent on occasions when I was out walking or driving with the participants. As I journeyed with Stan:

It was a lovers' lane up here,
on this side farther down.
Sometimes we was going, eleven o'clock from work,
we used to count the number of cars,
sometimes up to twelve, eleven.
Then this motorcycle cop starts coming into them:
little by little he cleared them all out. . .

And as we walked around the neighborhood Raymond was constantly reminded of the stores which had been, the people who had lived in the houses we passed, and events which had transpired within them. It was apparent that it was not the physical structures which mattered as much as the sense that something had happened in these places. Thus, as we passed the front yard of a corner home, Raymond was reminded of the news office which had once stood on the spot where roses now bloomed. As we passed the supermarket, Edward recalled the mansion which had been Thomas's palatial home.

People were important cues to geographical fantasy in two ways. First, as elements or objects of place they often provided the focus of its resurrection in consciousness. Second, the mere presence of people seemed to act as a stimulus to fantasy.

One afternoon Raymond took me on a journey through an environment of his past. As we sat at the table talking about the many jobs he had held, he began describing the route he had taken during his spell as a baker in the neighborhood (Figure IV.2). He began tracing the path on a scrap of paper, and as he drew he talked about his day. There were the people he knew very well on Durham Street. There was the woman whose dog had been particularly ferocious. He would save some of his best pastries for Mrs. Sargent, for she generally gave him coffee. He didn't hurry. There were people to talk to. He would time his route so that he would be passing his home in time for lunch. Then off again around the rest of his route. As he talked Raymond focused his descrip-

188

tion upon accounts of people and events in the homes he served. The people of his past acted as cues in the reconstitution of the space.

The longer he continued the more he seemed to become absorbed in his own private reverie. It did not seem to matter if I were present or not. Or did it? I had noticed on many previous visits that, like Evelyn and especially Marie, once he started talking about the past, he would ramble on and on, sometimes oblivious to my requests for clarification. More insistent probing seemed to provide merely an irritating distraction. In retrospect I suspect it was very important for me to be there. My presence acted as a cue: it legitimized his reflections. The presence of people who have shared the experience of a place may be particularly important. Sometimes when I was visiting Evelyn, her sister would arrive. Very often the conversation between the two would focus upon a sharing of the way things had been. The environments of Marie's past, too, were reactivated through interaction with her friendship network.

The *media* provided a fourth set of cues to geographical fantasy. A television weather forecast could stimulate involvement in an Arizona environment, reference to Japan could provoke reflection on a far away city. The newspaper also provided an important cue. Raymond, Marie, Stan, and Evelyn were particularly interested in the obituary section. Indeed, I discovered that Evelyn made a practice of cutting out and preserving the obituaries of her friends. This seemed to reflect a desire to keep informed about the deaths of peers rather than a morbid obsession. How often the passing of a close friend revived memories of a day spent together on a far away beach I could not tell.

INTEGRATION: A FRAMEWORK FOR INTERPRETING GEOGRAPHICAL EXPERIENCE

Thus far we have broken down the participants' geographical experience into four modalities. These have been considered as if they were independent. Clearly they are not. At this point it is necessary to merge action, orientation, feeling, and fantasy in developing a holistic framework for interpreting the totality of the individual's geographical experience.

A single environmental transaction involves an interdependency among all four modalities. When Stan strolled from his home to a bar, his experience could be viewed in terms of his pattern of action in traversing space. However, understanding of the trip is enhanced by an awareness of his orientation—the way he employed schemata in avoiding the cracks in the sidewalk, and the rationale behind his pausing a little longer before crossing a busy Easthill Street at noontime. A further layer of insight is added by acknowledging the influence of feelings about places. We begin to understand why he traveled to Steve's Bar and then on to Murphy's, bypassing a "plush" and "ritzy" Selena's. Finally, understanding of the trip is enriched by considering a range of fantasies it may evoke. The excursion is transformed into a voyage through a vibrant living landscape, a trek through space and time in which the legacies of intimate association over many years find resonance.

While at this stage it is not possible to explain the intricate details of interdependency, it is clear that in concert, actions, orientation, feelings, and fantasies, constitute a complex of modalities (summarized in Figure VII.5), which tends towards internal *consistency*. Stan's pattern of action in walking to the bar was consistent with his orientation within the local setting. Choosing the outside of the sidewalk was consistent with an awareness of the cracked paving closer to the fence. Pausing on Easthill Street was consistent with his anticipation of the traffic which he knew would make the space hazardous at lunchtime. Considering feelings, it would not have been consistent for his actions to include habitually traveling to a bar he felt was "plush" and "ritzy," or to the "places there they've got all the queers going." Similar consistency can be illustrated for all five participants.

The individual's geographical experience then, reveals an internal consistency. The modalities are the means through which the individual relates to a geographical lifespace. Although the modalities are shared, the pattern of geographical experience revealed by each participant is highly individualistic. To understand this diversity of expression, it is necessary to set geographical experience within a broader context: to view it as the outcome of an ongoing dialogue between the person and a geographical lifespace.

On the simplest level it is apparent that each participant has established a pattern of activity which reflects a balancing between personal capability and the opportunities provided by the environmental context. An *adjustment* is maintained which expresses the tension between these two interacting elements. However, as the vignettes and our previous discussion have indicated, the outcome, in terms of geographical experience, is far more than the expression of a simple reciprocal stimulus/response relationship. In each case the state of adjustment expresses a more complex transaction between the person as a total being and his geographical lifespace. In elaborating this notion, it is

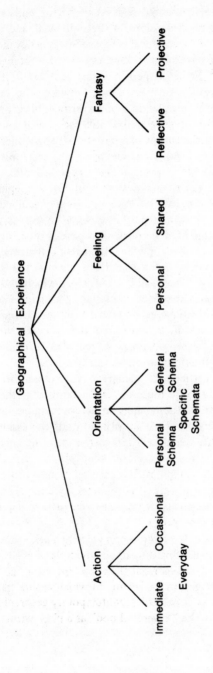

Figure VII.5 Geographical Experience: A Complex of Modalities

191

important to emphasize that our definition of person embraces both physical and psychological dimensions of the individual; and to reiterate that our definition of geographical lifespace involves not merely the contemporary physical environment, but rather "the totality of the experienced milieu."[21] For each person adjustment results from the effort to maintain harmony or consonance between who he is as a person (identity), and the manner in which, through his geographical experience, he relates to his geographical lifespace. Such consonance was apparent in Stan's maintenance of a pattern of geographical experience reflecting, on the one hand, awareness of his physical limitations, a "killing time" attitude, and overwhelming acceptance of a negative self-image; and on the other, an assessment of his geographical lifespace as hostile and limiting. In Marie's case, geographical experience expressed consonance between a person viewing herself as a feisty, community and family oriented guardian of a passing social order, and a complex geographical lifespace in which the harsh and alienating contemporary setting and vestiges of a friendly and supportive historical milieu were superimposed. Indeed, all five participants had established a pattern of geographical experience revealing consonance between personal identity and the unique geographical lifespace each inhabited.

At this point it is helpful to summarize the proposed theoretical framework graphically. Such a summary is provided in Figure VII.6. However, as it stands, this framework still does not fully account for the diversity of the participants' geographical experience. To accomplish this it is necessary to more directly acknowledge the role of individual personality and autobiography in determining the manner in which person/geographical lifespace adjustment finds expression. Autobiography and personality are closely interwoven determinants of individual adjustment, by virtue of the role they play in defining both elements of the person/geographical lifespace relationship. It is hardly necessary to elaborate on the role of autobiography and personality in defining the person. However, some comment on their role in defining geographical lifespace is appropriate.

Considering first the role of autobiography, the participants inhabited a geographical lifespace consisting of an ensemble of environments. Selective remembrance of scenes from their lives involving, in the process, a reconstitution of the environments in which events transpired, meant that each individual inhabited a contemporary geographical lifespace which was unique. The "neighborhood" as a place within Marie's

---

[21]See Chapter II

192

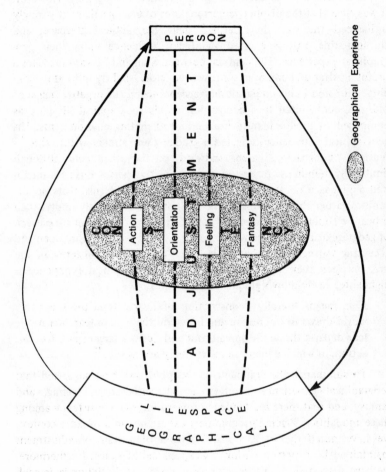

Figure VII.6 A Context for Geographical Experience

193

geographical lifespace was a fusion of the threatening, alienating space of the present, and a more benign space of the past. It was not the environment I could observe, but rather a complex synthesis of both a present and a past place. Her geographical experience expressed adjustment in relation to this composite "neighborhood" of her personal history.

It is difficult to separate personality from autobiography. However, it was clear that the distinctive personalities of each participant strongly influenced the way they defined their geographical lifespace, and through this process exerted significant influence upon their geographical experience. The contrast between Stan and Edward provides a useful illustration here. Stan's introverted, stoic, slightly morbid personality, mirrored in his unquestioning acceptance of the negative image of old age, contributed to a conception of his geographical lifespace as composed of predominantly hostile and alienating environments. His geographical experience reflected a grudging adjustment within this essentially "constricting" arena. Edward, on the other hand, although limited by blindness, possessed a more positive personality. He inhabited a lifespace constituted more optimistically in terms of its opportunities rather than its constraints. If Stan's glass was half empty, then Edward's could be characterized as half full. The richness of his pattern of geographical experience was in part a reflection of his more open and accepting attitude. The contrast between the two men in terms of the link between their personalities and their geographical experience is highlighted in Bollnow's eloquent observation:

> *Fear* means literally "constriction of heart," and the outer world draws in oppressive and heavy on the man in fear. When fear departs the world spreads out and opens a larger space for action, in which a man can move freely and easily.[22]

To summarize the argument to this point, we have suggested that geographical experience involves action, orientation, feeling, and fantasy, and that there is a tendency toward internal consistency among these modalities. Placing geographical experience in a broader context we have noted that it is a manifestation of a state of adjustment maintained by a person within a geographical lifespace. Furthermore, we have suggested that the precise form of this manifestation is, in each case, dependent on the unique autobiography and personality of the individual. At this juncture, in order to round out the framework and

[22]Otto Bollnow, *op. cit.*, 1967, p. 185.

bring it to life, it is helpful to introduce a dynamic component by considering the model as a system in process.

Through time the nature of individual adjustment changes in response to changes in both the person and his geographical lifespace. This ongoing feedback process is represented in Figure VII.6. Changes in the geographical lifespace may result from physical environmental change, such as the erection of the cemetery fence which prevented Edward from strolling in the area he formerly frequented beyond the highway, or the demise of the neighborhood stores Evelyn had patronized. Other more subtle changes are more directly linked to the process of the person's life. The acquisition of new information via the media, the relocation of a son to Tokyo, or an Arizona vacation, can result in elaboration or reconstitution of the individual's geographical lifespace. With the passing of time the individual also changes. Physical capabilities change, and as the individual ages, personality, values, and self-conception undergo modification. The person may become morose and withdrawn like Stan; defensive and defiant like Marie; quietly accepting like Edward; indeed, the range of changes is almost infinite. As personal status and geographical lifespace change so also does the nature of the adjustment the individual maintains within a total context.

Change on this level is mirrored by changes in geographical experience. Patterns of action, established and maintained across the span of life, are modified in accordance with changed circumstances. Orientation also changes through time as the mirror of personal history. For example, schemata may become progressively more refined and subtly differentiated as familiarity with environmental settings makes them increasingly implicit expressions of geographical experience. Feelings about places also undergo transition. Familiar settings may evoke an ever richer array of emotional identifications as the imprint of events, which may have transpired many years previously, merges with affective affiliations of more recent origin. Some places may seem to become more friendly and supportive; others become increasingly sinister and threatening. And finally, involvements in the environments of fantasy evolve in accordance with autobiography in such a manner as to maintain consonance between the person and the geographical lifespace he or she inhabits. Thus as Marie had grown older, fantasy involvement in the friendly spaces of neighborhood past assumed increasing importance in the maintenance of a state of adjustment within the Winchester Street context.

195

In sum, we may view geographical experience as the expression of a constantly evolving person/geographical lifespace system in dynamic equilibrium. At any one time the state of adjustment maintained by the person is, at the same time, an expression of who he is, where he has been, and even where he would wish to be.

From critical appraisal of five elderly persons a framework has been proposed which interprets their geographical experience as a highly individualistic meshing of four overlapping modalities. As time passes and person/geographical lifespace adjustment changes, there is a transition in the quality of geographical experience. But this framework could be applied not only to the participants but to any individual, regardless of age. The tantalizing question of age-related changes remains unanswered. In the concluding section of this chapter consistent developmental changes accompanying the aging process are postulated. Because of the limited size of the study population, the relatively brief duration of the research which precluded longitudinal comparisons, and the necessarily speculative nature of the inferences, the conclusions are presented in the form of a hypothesis deserving of serious consideration.

A HYPOTHESIS OF CHANGING EMPHASIS

As the individual grows older there are characteristic changes of emphasis both *within* and *among* the modalities of geographical experience. These changes involve *constriction, selective intensification,* and *expansion.* Constriction refers to changes tending towards geographical lifespace limitation and closure: it is primarily manifest in the realm of action. Selective intensification, involving increasing psychological investment within particular places, is postulated as the dominant motif of change within the realms of orientation and feeling. Finally, expansion refers to changes in which the geographical experience of the older person is enriched. It is suggested that, with advancing age, there is an expansion of the role played by fantasy in the totality of geographical experience. In elaborating these notions we turn first to explication of changes posited to occur within each of the four modalities.

In the domain of action the predominant process is one of constriction. Immediate movement is limited by reduced agility stemming from physiological decline. This change is reflected in curtailment of the range and diversity of everyday activity, as the older person accommodates to decreasing potential for environmental negotiation. However, an interesting paradox appears when occasional trips are considered. The participants' experience supports a hypothesis that, at least until advanced old age, there is an expansion of activity in this domain. The freedom afforded by retirement and a healthy "young" old age may temporarily "liberate" the older person by facilitating travels to vacation, or to visit relatives and friends in geographically distant locations.

It is necessary to be more tentative in postulating changes in orientation. Changes are closely linked to changing patterns of action. Considering first, changes relating to the personal schema, with advancing age there is an increasing tendency for unsteadiness and disorientation. It can be hypothesized that there is an increase in the importance of certain environmental supports for maintaining balance and orientation. Thus Stan often used a cane. He would clutch my arm as we crossed the street, and would select paths providing him with something to grab onto, should he feel himself falling.

Reduced capabilities also lead to changes in specific schemata. It is hypothesized that there is a selective intensification in the importance of particular environmental cues in traversing space. This involves increased sensitivity to more subtle environmental nuances. Edward, for example, had become sensitive to the slight uphill gradient and the gusty breeze he would face if he elected to walk home from the bus stop at the "Four Corners" in preference to the one on Imperial Street. All the participants were more cognizant of curbs, steps, and other small barriers which had not bothered them when they were younger. In this context, familiarity with a path becomes increasingly important. Familiarity facilitates gradual adjustment to reduced physiological and psychological capability, by making it possible for the older person, over time, to reorient progressively through developing specific schemata in which guiding landmarks and potential hazards are highlighted. The process involves "selective attention" to certain environmental features which both absolutely and relatively intensifies their perceived importance. Coming to "know" the route in considerable detail through this process enables the older person to continue negotiating a path successfully.

It is hypothesized that there is also a selective intensification of the importance of certain zones within the general schema. Figure VII.7 was compiled from my subjective assessment of changes in each of the participant's general schema. Clearly this material can only be viewed as crudely suggestive. You are encouraged to assess the inferences which have been made in terms of your own reading of the five vignettes. There was some variability. Nonetheless, it is hypothesized that, as the individual grows older, proximate zones—the home, and especially the

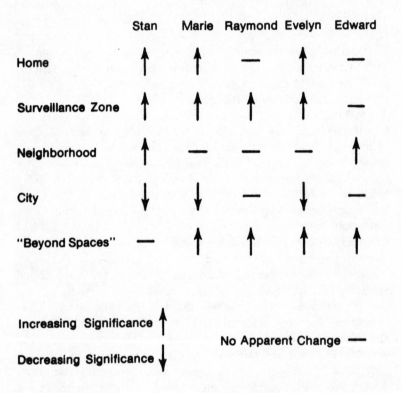

Figure VII.7 The Changing Emphasis of Zones within the General Schema

198

surveillance zone—become more significant components of the general schema. Home had always been an important domain for Marie and Evelyn, but as they had grown older they acknowledged that it had become even more central to their lives. As environmental constraints became more limiting in relation to his deteriorating health, and he was obliged to spend more time at home, Stan's abode assumed greater prominence within his general schema. With regard to the surveillance zone, the suggestive evidence provided by the research is strong. For all the participants, with for obvious reasons the exception of Edward, the more frequent watchful monitoring of events outside the home enhanced the importance of this narrow band of space. In addition, it is hypothesized that the significance of selected "beyond spaces" is intensified. The great concern of four of the five participants with the geographical settings in which their children or relatives resided, provides some support for this suggestion.

Feelings about places are intimately linked with the events of the individual's unique biography. However, the research provided some indication of a tendency towards a selective intensification of affective bonds with place with advancing years. Once again, there was considerable variation among the participants, but three general trends are postulated. First, paralleling changing emphasis in the general schema, there is increasing affective identification with the proximate physical setting, and with "beyond spaces" where children reside. The home, surveillance zone, and to a lesser extent, the neighborhood, had over the years become progressively more richly imbued with emotional significances. In addition, it was clear that Marie, Raymond, and Evelyn had developed strong feelings for the far away spaces in which their children lived. Such an intensification of emotional attachment, it is postulated, enhances a sense of security and belonging.

Second, it is hypothesized that there is a selective intensification of feelings for locations of special significance in the older person's past. Thus for Edward, the area he had fearfully hurried through on the dark evening of a Jekyll and Hyde performance; the railroad crossing, scene of so many tragedies; the cemetery, rich landscape of a boyhood playground: even the British Men's Club he visited once in a while, had in each case, come to evoke a stable array of feelings. For Marie, the area around Duvalle Square, memorial to her son; the hall where she had danced in a new red velvet dress on the memorable evening of her twenty-fifth wedding anniversary; the home next door where she saved a life; and even the Quebec village of her childhood, had all become affectively fixed, indelibly imprinted with a cherished set of emotional meanings.

Turning from personal to shared feelings, it is suggested that there is an intensification of a communal sense of the neighborhood's social aura. Continuing peer group reaffirmation of a "field of care," the tightly knit interdependent community which had existed, seemed especially important to the French participants. By constantly exchanging reminiscences they were able to preserve and intensify their affinity with neighborhood space.

Selective intensification of feelings about spaces may be far more than merely the coincidental outcome of lengthy residence in a single setting. It is postulated that it represents a universal strategy employed by older people to facilitate maintaining a sense of identity within a changing environment. In this context, the survival of older people's peer group social networks may be extremely important.

Finally, the research process furnished evidence for a hypothesis that as the individual grows older, there is an expansion of the role played by fantasy within the totality of geographical experience. It is suggested that both reflective and projective geographical fantasy assume increasing importance in older people's lives.

Regarding reflective fantasy, Edward freely acknowledged that he spent more time "dreaming":

And sometimes I think about places I went. I'm the type of person that likes to dream a little. Dream, I call it dreaming. I think it's good for you.

And as Marie had grown older it was clear, both from the reports of her friends, and merely being in her company, that she retreated into reverie more frequently. This did not constitute a pathological decline: in part, it was a natural response to the perceived disintegration of the world in which she found herself. But beyond this, she seemed to derive great pleasure from immersing herself in the environments of her past, and sharing her memories with those who would take the time to listen.

There is some support for the notion of expansion in the role of reflective geographical fantasy in research suggesting the importance of reminiscence in old age.[23] In their study Lieberman and Falk found

[23]Robert N. Butler, "The Life Review: An Interpretation of Reminiscence in the Aged," *Psychiatry,* XXVI, 1963, pp. 65-76; Arthur W. McMahon and Paul J. Rhudick, "Reminiscing," *Archives of General Psychiatry,* X, 1964, pp. 292-298; Arthur W. McMahon and Paul J. Rhudick, "Reminiscing in the Aged: An Adaptational Response," in Sidney Levin and Ralph J. Kahana, eds., *Psychodynamic*

that: "The aged were considerably more involved or interested in reminiscence activity than were the middle aged."[24] Butler has suggested there may be a distinctive reminiscence phase through which older people pass in seeking to integrate their lives. Lewis posits that "reminiscing and identifying with one's past may be a defense mechanism for certain older people."[25] And McMahon and Rhudick found that older men who reminisced seemed less depressed when compared to non-reminiscers. Thus it is postulated that increasing concern with reflective fantasy, and by extension with geographical settings of the past, is not only a distinctive developmental trend but is also psychologically beneficial.

Raymond provides the best illustration for elaboration upon a hypothesized expansion in the role of projective geographical fantasy. The special clock on his mantle, preserved letters and postcards, and prominently displayed photographs of his children, were cues to involvement in the worlds of his family which vicariously transported him across great distances into far away settings. It is suggested that Raymond's involvement in his Arkansas garden, and interest in his son's Tokyo apartment is not atypical. Rather it illustrates a form of experience common to many older people. In his own words, "Your success after you've raised them is *to watch their family grow and take part of it.*" Involvement in the worlds of children and living through their experience seems to be an implicitly accepted role of the older person in our society.[26] As societal changes have resulted in the tendency for increasing geographical separation of families, less direct communication processes such as the telephone and letters, together with the media (for example, Evelyn's watching the weather forecast in

Studies on Aging: Creativity, Reminiscence and Dying, New York: International Universities Press, Inc., 1967, pp. 64-78; Charles N. Lewis, "Reminiscing and Self-Concept in Old Age," *Journal of Gerontology,* XXVI:2, 1971, pp. 240-243; Morton A. Lieberman and Jacqueline M. Falk, "The Remembered Past as a Source of Data for Research on the Life Cycle," *Human Development,* XIV, 1971, pp. 132-141.

[24]Morton A. Lieberman and Jacqueline M. Falk, *op. cit.,* 1971, p. 139.

[25]Charles N. Lewis, *op. cit.,* 1971, p. 243.

[26]There is of course an important question here, concerning the extent to which this phenomenon is a function of old age rather than family status: it can be argued that the observation applies equally to many middle-aged persons. As this study was concerned only with elderly persons, consideration of this question was not possible. However, this issue clearly requires addressing in future research.

Arizona), have assumed increasing significance. In addition, involvement in the worlds of children increasingly assumes a vicarious rather than direct form. The implications for older people's geographical experience include a tendency for increasing involvement in fantasy with events in far distant locations.

Considered independently, hypotheses concerning age-related changes in the four modalities of geographical experience provide an array of potentially fruitful avenues of inquiry. Viewed in conjunction, these seemingly disparate propositions can be welded into a coherent hypothesis concerning the manner in which geographical experience changes as the person grows older. This hypothesis suggests that there is a developmental consistency in the changing adjustment maintained by the person within his geographical lifespace.

As a major proposition stemming from this inquiry it is hypothesized that *as the person grows older there is a change of emphasis within geographical experience involving a constriction in the realm of action which is accompanied by an expansion of the role of geographical fantasy.* A reduced level of physical participation within a local contemporary setting is, at least in part, compensated by increasing vicarious involvement in environments of the imagination which may be spatially and/or temporally far removed. The process is a logical consequence of the older person's reduced capability to negotiate the physical environment, a tendency toward increasing emphasis upon reminiscence, and a propensity for increasing concern with events in geographically remote environmental settings.

As this change of emphasis between action and fantasy occurs, it is hypothesized that *there are consistent accompanying changes in the older person's orientation within space and in feelings about the places of his life reflecting a selective intensification of involvement.* Specific schemata become differentiated in terms of more refined environmental cues. There is also a selective intensification in the importance of proximate zones (especially the home and surveillance zone) and certain "beyond spaces" (notably the homes of children) within the general schema. Associated changes in feelings about place include an intensification of affective bonds with proximate spaces and locations where children reside, and a selective intensification of emotional identification with places of importance within the individual's personal history. The total complex of geographical experience evolves in a consistent and coherent manner.

So, from the complexity and confusion of the data we begin to discern some pattern. By abstracting from individual experience it has

202

been possible to trace the outlines of a dynamic framework for interpreting the changing geographical experience of the older person. Clearly, there are many unanswered questions, gaps in the framework, even some contradictions. But at this stage such inconsistencies provide a challenge rather than cause for alarm. For the exploration has merely begun!

CHAPTER VIII

WIDENING HORIZONS

You progress not through improving what has been done, but reaching toward what has yet to be done.

Gibran

The search for understanding is an ongoing process, the never ending quest for deeper insight. Periodically it is helpful to pause and review progress, to isolate areas where further amplification is required, and to speculate on the potential application of emerging ideas. However, appraisal of the substantive conclusions cannot be divorced from assessment of the process whereby insights were derived.

OLD IS ANOTHER COUNTRY

I sought to venture unencumbered into another world in the hope of returning wide-eyed and bushy tailed to tell tales of exciting discoveries in alien territories. The previous pages provide an indication that to some degree I was able to accomplish this. But a few observations on the journey may aid future explorers. For beyond the expected tribulations of any such exploration, several problems were encountered which can be attributed to the route I chose.

A focal issue is the degree to which it was possible to facilitate sensitive insight by establishing dialogue with the participants. To claim total success would be delusional. Perhaps because of my sex, it seemed I was able to develop a closer relationship with the male participants. Edward had no apparent qualms about revealing a quietly bitter resentment of his sister's rigid insistence on retaining a less than perfect housekeeper. Stan would talk without embarrassment about his illness and impending death as he shared his heavy sadness. Raymond was never reluctant to confide uncharitable sentiments about some of his neighbors. Many intimate and poignant moments were shared. On the surface, Marie and Evelyn seemed just as friendly and open. Yet I

205

sensed they held more in reserve. There was an unyielding wariness of confiding innermost feelings, a reluctance to reveal any sense of vulnerability to a mere "boy." As Evelyn remarked on one occasion, much to my chagrin: "You, you're young. You look like a child."

As expected, particularly during the early stages, there was much fumbling and role playing. The participants did not know how to react to me. With the exception of Edward, they seemed socialized to awed respect and yet, at the same time, a skeptical mistrust of "college people." It was as if they accepted at the outset that I would not wish to know them for themselves, but rather as objects, statistical entities, mere pawns in some incomprehensible game. It took several weeks to accept that I really wished to know them as people—in all their complexity. The barrier seemed to dissolve as our relationships became more fully reciprocal. There was increasing interest in my own life and work. Evelyn would ply me with questions about my past, Edward seemed more interested in my future aspirations, and Raymond would constantly inquire about my wife and home life. Gradually each participant overcame initial reluctance to call me at home, to cancel meetings, to invite me over to play cards or watch television, or to take advantage of my offer of transportation. As I became more trusted, the tenseness disappeared, and we were more fully able to share, and to mutually explore facets of geographical experience.

The main problems were not those experienced by the participants but were a function of my own involvement. It was difficult to transcend the influence of my academic socialization, especially during the early weeks. I found myself asking leading questions, probing for support of preconceived notions. It was difficult to refrain from directing conversations. The essence of the problem was an inability to suspend judgment temporarily. It was exceedingly hard to divorce myself from the constraining awareness of conventional arguments about the way older people are generally considered to experience space. I fought for liberation from the cumbersome lexicon of "activity spaces," "distance decay functions," "mental maps," and other blinders of established social science lore. In retrospect I realize I was never fully able to break free from these constraints.

As I became more successful in suspending prejudgment, other problems, stemming from my ambivalent existence, became apparent. When I was with the participants it was as if there was a separate reality. I began to talk and even to think within a different framework. I would use Stan's phrases, his terse style of interacting with his barroom friends. I became so deeply immersed within the participants' worlds that it became difficult to fulfill obligations outside my research without retreating from the level of rapport I was establishing. Paradoxically, my involvement started to become counterintuitive. It became more difficult to pull back from the experience, to think critically and to interpret. I found myself grappling with a dilemma previously conceived of in merely abstract intellectual terms; the issue of establishing balance between a level of commitment necessary to let sensitive and grounded insight evolve and one facilitating the communication of such understanding.

Standing by a grave, sharing quiet reminiscences with an elderly man you have come to know, yields a special kind of interpersonal knowing. The phrases, the tone of voice, even the pauses and periods of silence, all are revealing. I could not *know* Edward's experience but I developed an awareness of his involvement with that place, and its meaning, which transcended mere knowledge that he had once played there. The difficulty arose in attempting to communicate this sense without reducing it to sentimental banality. Ability to accomplish this is highly dependent on writing skills: it is necessary for the social scientist to become poet. And herein lies the rub, for few are so gifted. In agonizing over appropriate phrasing, in the drafting and redrafting, in the frustration stemming from clumsy efforts to convey mood, and in the process of attempting to project a sense of the experience, the problem of communication was clearly revealed. Problems of accurate representation are common to all inquiry; but they are accentuated when the emphasis is upon subtle nuance. There are no models for writing, no rules to be followed apart from the constant striving for authenticity.

In describing experience it is necessary to make inferences about the meaning of events and to be selective in incorporating material. Acknowledging this, how accurately do my vignettes portray the participants? In Marie's case, for example, a characterization revealing her as a defiant, defensive, self-reinforcing person, inhabiting a fantasy world, is only one of many possible interpretations. On one level the issue may be posed as one of the degree to which my interpretation mirrors Marie's existential relationship with her environment, rather than my externally imposed rationale. In an ultimate sense it can only be the latter. However, inasmuch as the research process resulted in an authentic relationship, Marie's experience is infused within my understanding and imbues the interpretation with greater sensitivity.

The real issue, of course, is whether my approach resulted in insights which enhance our understanding of older people's geographical experience. There are two questions here. To what extent did the

207

methodology facilitate insights which could not have been gleaned in any other manner? One answer to this question is that *de facto* the approach has resulted in perspectives which have not been derived from alternative methods. Many of these insights can be directly attributed to intensive involvement with participants. Only such intimacy could furnish an understanding of Raymond's attachment to his Arkansas garden, reveal the importance of gaps in fences within Edward's geographical experience, or provide entry into the colorful spaces of Marie's neighborhood past. The second question is: Has the research resulted in insights which are useful? This can only be assessed in the light of future attempts to verify the ideas developed. In the interim, potential usefulness must be judged by you the reader.

ON THE GEOGRAPHICAL EXPERIENCE
OF OLDER PEOPLE

Amplifications

Having commented on the route, we turn next to consideration of the substantive objective, in attempting to provide some perspective on the territory I explored. The expanded conception of the older person's geographical experience presented in this study conflicts with the prevalent image of old age as inexorable closure—a process of progressive spatial constriction. In particular, by acknowledging the role of geographical fantasy within a more holistic interpretation, we see that physical limitation does not necessarily imply psychological retreat. Indeed, it may signify the reverse. We can begin to detect a rationale behind the observed propensity of older people to muse on environments of their past, and to surround themselves with cues linking them with the worlds of their children. It becomes possible to understand how many older people are able to lead fulfilling lives in the most squalid surroundings. The proximate contemporary setting is only a part of their geographical lifespace. Indeed, in this acknowledgment may lie a potent explanation of the reason why morale/environment correlations are often so ambiguous. Many such studies only consider the contemporary characteristics of housing and environment.

We can begin to explore overlap with a growing body of literature in psychology which, focusing on older people's increasing interiority, propensity for reminiscence, and concern with the integration of their life experiences, is providing the outlines of a perspective on aged status

208

as a developmentally distinctive phase of the lifecycle. The hypothesized transition in the quality of geographical experience may constitute an important theme within this perspective.

In pursuing this exciting possibility it will be important to establish the degree to which changing emphasis in geographical experience is a distinctive feature of growing old, rather than merely a culturally conditioned response to the constraining circumstances under which older people in American society are obliged to live. Gerontological research has already once fallen into this ethnocentric trap in at first embracing and accepting the universality of the disengagement theory of aging.[1] However, before we can even usefully contemplate probing this question, the proposed theoretical framework requires further elaboration and empirical testing. It is helpful to consider some directions we might profitably pursue to this end.

Does a dynamic framework incorporating action, orientation, feeling, and fantasy provide a coherent expression of the older person's geographical experience? Within each modality there is a need to explore overlap among the various subdimensions which have been distinguished. How much was the pattern of Stan's daily excursions (everyday activity) constrained by the difficulties he experienced in climbing stairs (immediate movement)? To what extent was his general schema in the zones adjacent to his home distorted by the motif of the specific schemata he employed in traversing frequently traveled paths? How closely did his personal feelings about Selena's Bar and "places there they've got all the queers going," reflect feelings shared and molded through interaction with his barroom companions?

It would also be helpful to further explicate linkages among the four modalities. In what ways were Marie's actions conditioned by the schemata she employed in orienting herself within the Winchester Street environment? Can we specify more precisely the way in which her feelings about the space, a complex synthesis of fear and affinity, impinged upon her pattern of action? To what extent could her actions be accounted for in terms of the rich fantasy spaces of neighborhood past she inhabited? Clearly there are as many questions as linkages.

Considering the framework on a more general level an array of insights can be anticipated from further probing the manner in which actions, orientation, feeling, and fantasy are incorporated within a person/geographical lifespace framework. How long does it take for an

[1] For a critique pointing out this error see, Arnold M. Rose, "A Current Theoretical Issue in Social Gerontology," *The Gerontologist*, IV:1, 1964, pp. 46-50.

individual to establish the intricate and finely balanced state of adjustment maintained by Marie or Raymond, indeed, each of the participants in this study? How important are personality and autobiography in determining the nature of adjustment, and hence the person's geographical experience? Elaboration of this extremely complex relationship would be facilitated by a detailed study of a single person's adjustment within his or her geographical lifespace, focusing specifically on relating geographical experience to biography and personality. What are at present shadowy and indistinct relationships must be brought into sharper focus.

The most exciting findings are often the most tenuous, for they tend to pose challenging questions for future exploration. So it is with the dynamic aspects of the proposed framework. It is especially important to seek refinement of the change of emphasis hypothesis. Is there a developmental consistency in the changing geographical experience of the older person? Does the hypothesis describe a universal process, applicable to all older people; or does it apply only to those who become physically restricted? In fact, is the phenomenon limited solely to old age? "Dreaming" is clearly not the exclusive preserve of the elderly. Most of us derive pleasure from the periodic contemplation of the places of our past. Does this propensity increase with advancing years as the reservoir of possibilities increases? Obviously there is a need for both cross sectional and longitudinal study of age related change in vicarious participation in environments.

How closely can the hypothesis be tied in with contemporary work probing the role of reminiscence in old age? As we have noted, much of this work provides support for a hypothesis of increasing emphasis upon vicarious experience with advancing age. However, there is also negative evidence. Giambra, for example, found no evidence of an increasing propensity for "daydreaming" among the elderly.[2] Is there steady expansion in the role of geographical fantasy with advancing age,

[2]Leonard M. Giambra, "Daydreaming About the Past: The Time Setting of Spontaneous Thought Intrusions," *The Gerontologist,* XVII:1, 1977, pp. 35-38, and, "Daydreaming across the Life Span: Late Adolescent to Senior Citizen," *Aging and Human Development,* V:2, 1974, pp. 115-140. For additional examples of work probing this important theme, see, Peter G. Coleman, "Measuring Reminiscence Characteristics from Conversation as Adaptive Features of Old Age," *Aging and Human Development,* V:3, 1974, pp. 281-294; Charles N. Lewis, "Reminiscing and Self Concept in Old Age," *Journal of Gerontology,* XXVI:2, 1971, pp. 240-243; Morton A. Lieberman and Jacqueline M. Falk, "The Remembered Past as a Source of Data for Research on the Life Cycle," *Human Development,* XIV, 1971, pp. 132-141.

or is there eventually a phase of decline, equatable with completion of the life review or imminent death?

Most important, illuminating insights may be anticipated from monitoring geographical manifestations of increasing vicarious environmental participation. There is a need to inventory, and to search for consistent patterns in the ensemble of displaced environments inhabited by each older person. And in the context of such amplification, it is necessary to probe the impact of changing propensity for vicarious experience of environments upon the older person's orientation and feeling. How does the person's orientation within space and feelings about the places of his or her life change in association with a changing emphasis between action and fantasy?

We can outline a rich prospectus for future elaboration of the notions emerging from this study. Such refinement must, however, proceed hand in hand with efforts directed towards substantiation. The present exploration has involved an extremely small population drawn from a common physical setting. As with all inductive inquiry, the insights derived can only be intuitively rather than formally generalized beyond the five individuals upon whom the findings are based. Such a sobering observation is not grounds for pessimism as, from the outset, the intention was never to claim universal validity. It was to generate experientially grounded ideas rather than to test *a priori* notions. The quest for more general substantiation belongs to the next phase of inquiry. Each of the ideas developed must be carefully tested employing statistically meaningful samples of diverse groups of older people. But if the findings are confirmed, what exactly would they mean?

Applications

It would be premature to explore in detail the implications of notions which at this stage are necessarily speculative. However, cursory sampling of a few possibilities illustrates potential applications of the ideas. In planning for the elderly, consideration of geographical experience has focused almost exclusively on the theme of compensating for progressive limitation of action. This has been manifest in programmatic emphasis on a variety of direct transportation services, and concern with creating physically "barrier free" environments. If the findings of this research are confirmed there emerges a rich potential for refocusing such strategies to facilitate a more refined client-centered response to the mobility dilemma. For example, one observed response to the limitation of everyday activity was increased reliance upon "sur-

211

rogate" trips by friends and relatives. We have services which facilitate transporting older people to resources, and programs (such as "Meals on Wheels") which bring resources to older people. But such options do not provide the flexibility to cater to what may be very important preferences of the older person (for example, Stan's use of Steve to obtain fish from a market on the far side of town). Such options would become more feasible with the introduction of a personalized courier or "surrogate trip" service. The emphasis here is on the "qualitative" aspects of such a service rather than its mere presence. The employment of local *facilitators,* familiar neighborhood residents who would be compensated for providing such a service, would effectively personalize such an effort to compensate for the physical constrictions confronting the older person.

Acknowledging the significance of occasional trips in older people's lives indicates a second set of programmatic possibilities. Church clubs and social agencies sometimes organize vacation trips for older people. However, such excursions do not provide sufficient locational flexibility in the choice of destination. Such flexibility would be an integral component of a support program to facilitate occasional trips to visit children or to travel to far distant places of the person's past. In the context of the suggested importance of geographical fantasy for older people such support would be far from frivolous.

The main programmatic possibilities suggested by this study stem from an expanded definition of geographical experience. Accessibility comes to assume more subtle dimensions involving psychological as well as physical definition. Acknowledging the role of specific schemata within orientation refines the issue of creating barrier free environments. We can recognize the need to identify sensory cues— visual, auditory, and tactile—which facilitate making the environment cognitively as well as physically traversable for the individual with diminished capabilities.

The observed importance of geographical fantasy indicates a potential for programs which would enhance the process of vicarious involvement in displaced environments. The provision of telephones to all older people who desired them would provide a mechanism for ongoing contact with such environments. Moreover, subsidizing certain long distance calls to children and other relatives would facilitate continuing involvement with important worlds beyond the confines of the neighborhood. "Your success after you've raised them is to watch their family grow and take part of it." Other possibilities for sustaining the richness of geographical fantasy exist on a social level. Could we not contribute to the pleasure of reflective geographical fantasy by encouraging older people, in both formal and informal contexts, to savor and exchange reminiscences on the places of their shared past? The burgeoning domain of continuing education for older people provides one practical possibility here. I have found that courses in "Living History" and "Autobiographical Writing" in which participants are provided with support and encouragement to immerse themselves within temporally and spatially displaced environments of their lives provide a particularly effective medium. In particular such options seem to "legitimize" geographical fantasy.

A second domain of potential application of the ideas emerging from this study lies in the area of design. In the location, site planning, and design of both community housing facilities and institutional living spaces we can begin to incorporate consideration of more subtle components of the older person's orientation within, and feelings for space. Designers are already beginning to acknowledge the significance of the surveillance zone (a realm within the general schema) in recognizing older residents' preference for designs which offer potential for monitoring scenes of bustling activity instead of the dull tranquility of bucolic vistas. We might enrich the lives of even the most physically restricted, by creating surveillance zones through the judicious design of "lowered" window sills to provide a view of the world outside from a sitting position, by placing beds by windows, or by facilitating visual access to the bustle of corridor activity. There would also seem to be a rich potential for innovative design strategies which would be conducive to the development of feelings for space. Thus far, efforts to facilitate the personalization of space have focused upon the differentiation of space within housing and institutions through prudent interior design (for example, the careful use of color), and upon providing space for the storage and display of treasured personal possessions. The latter strategy also provides for maintaining cues to geographical fantasy (photographs, artifacts, and other memorabilia).

This observation can be linked with a third area in which this study may have important application, the issue of relocation. One hardly surprising finding has been a confirmation of the strength of many older persons' attachment to a neighborhood milieu. Many older people experience severe (on occasion fatal) traumas during relocation from such settings to new housing and especially to institutional settings. The stress of severance is generally assumed to be primarily a problem of adjustment to a new environment. However, recent research has indicated that the characteristics of the new setting may not be as important in determining successful relocation as has been traditionally believed.[3] Many older people experience stressful reactions *before* entering the new environment. Indeed, the problem may be effectively recast as one, not so much of facilitating adjustment to the new, as of easing the process of severance from the old. How does this study contribute to the solution of this dilemma? Essentially we are dealing with a situation where the older person may be grieving for a lost geographical lifespace within which an intricate and intimate adjustment (involving action, orientation, feeling, and fantasy) was established. The need is for strategies which ease the process of relocation by facilitating continuing vicarious involvement within the old milieu, even though physical severance has occurred. The transfer of cues to geographical fantasy, such as photographs, mementoes, and other significant artifacts, into the new setting provides one means of sustaining such involvement. In addition, the sanctioning, indeed encouragement of geographical fantasy (both reflective and projective) by enlightened staff may prove more effective than the often dubious benefits of "reality" therapy in facilitating adjustment to the new setting.

A final area in which the research reported in the previous pages has relevance is perhaps the most important. This entails an enhanced sensitization of professional human service practitioners to a dimension of older people's experience which has been hitherto ignored, or at best misunderstood. Hopefully, the vignettes and the interpretation stemming from them will provide the stimulus for improved understanding of older people's relationship with their environment. The shabby eccentric client, seemingly oblivious to the squalor of a run down tenement he calls home, may, in fact, have established an extremely complex, subtle, and resilient adjustment within his geographical lifespace. Actions may be finely attuned to environmental and personal potential. Orientation may be honed to the point where movement becomes independent of capacity for precise sensory discrimination. Long experience may have imbued the space with a rich reservoir of meanings. Finally, experientially, the person may be "inhabiting" different times and places. This is not to suggest the sanctioning of continued residence in the appalling living conditions under which many older persons are forced to exist. Rather, the intention is to foster more sophisticated understanding of the kinds of coping and compensating strategies which are employed as the older person adjusts to his or her environmental context. I believe that deeper understanding facilitates better helping.

[3]Sheldon S. Tobin and Morton A. Lieberman, *Last Home for the Aged,* San Francisco: Jossey-Bass, 1976.

STANDING BACK

At this point it is helpful to step back and attempt to view the study in broader perspective. Irrespective of history's verdict on the framework generated through this inquiry, the research process has served a useful function in reaffirming the richness of older people's lives and the falsity of many societal stereotypes. At the risk of appearing overly polemic it seems appropriate to conclude with some assessment on this level.

My exploration provided reaffirmation of the diversity and creativity of older people. In Stan's stoic resignation, Marie's aggressive defiance, Raymond's jovial acceptance, Evelyn's placid equanimity, and Edward's calm accommodation, we see five highly individualistic responses to growing old. Physical capabilities differ, support systems contrast, and aspirations, values, and dreams reveal great diversity. Each participant is a vibrant unique human being who has displayed great resilience in adapting to changing circumstances. It is hard to think of them as members of a common class, "old people." There is a need to turn away from a tendency to *homogenize*. Older people are not all the dynamic outgoing individuals parodied by social optimists. Neither are they the placid disengaging folk portrayed in the prevailing stereotype. Some are friendly, open, ebullient; others are reclusive, gnarled, and bitter. Most, like the participants in this study, are a subtle combination of diverse personal hues. This diversity is clearly reflected in their geographical experience. A closing lifespace perspective seems, in retrospect, to be a cruel and myopic parody.

Closely related to the tendency to homogenize is the *"be like us syndrome."* It seems that those of us who are not elderly are only prepared to relate to older people able to conform to our own modes of being. Older people have lived in different times, often hold different values, and experience environments differently, both because of their history and changes associated with their aging. What to the young is alienating isolation may not be stressful to older people who, like Edward as he sits "dreaming" in a broken down car, may relish the opportunity to reminisce in solitude. The most invidious manifestations of the "be like us syndrome" operate on the level of personal interaction. Why do we so often act as if fantasy and reminiscence were pathological? We shift uncomfortably, and avert our gaze when an older person begins to dwell on the past or drifts into musing about far away relatives. Can we not accept that for many older people it is important

215

to share the richness of their lives, and to transport us to the places they have frequented?

It is also necessary to strive for liberation from a pervasive *condescension* in our dealings with older people. At the commencement of my research I recall feeling sorry for Stan, seemingly condemned to last days of dull drudgery in a limiting environment. How sad that Marie was unable to face up to the realities of the deteriorating Winchester Street neighborhood. Poor Evelyn, living in her limited social/spatial world. If only she could have had the education, the experience, and the opportunities which I, child of technological society, possessed. A typical reaction, I suspect. As all-knowing modern men, all we see is the "sadness" of limited social horizons and the "parochialism" of strong attachment to place. We note the preserved artifacts and mementos, the photographs of children and the treasured scrapbooks. We listen with polite impatience, half hearing the words, barely participating in the seemingly unsophisticated conversation. And we are filled with an urge to protect. In many ways this is an expression of arrogance. It involves disregard for abilities which facilitated survival through world wars, economic depression, and testing personal crises. As my relationships with the participants developed, I became more and more in awe of the subtle sophistication, the intricate and yet resilient balance, of the niches they had carved for themselves within the spaces and places of their lives. For all our technological gadgetry and space age wisdom it is debatable whether we will live a richer old age. Yet condescension often provokes us to proffering unwanted "help." More tragically, benign concern can merge into subtle coercion, or worse, direct restriction "in their best interests" sometimes verging on totalitarianism. When did Stan give up his right to decide who could converse with him on his death bed?

In sum, there is a need to break free from prevailing social attitudes, which have served to alienate the elderly and to instill within us a view of their lives as ones of inevitable spatial withdrawal. Such liberation will involve neither maudlin sentimentality nor anguished wringing of hands, but a realism based on authentic relationships in which the beauty and the blemishes, the constrictions and new freedoms, and the joys and the sorrows of old age are openly acknowledged. Herein, I believe, lies the key to understanding how Marie, as she reviews her scrapbooks, as she muses on her daughter in Florida, and as she reminisces on the neighborhood of her past, though she remains in a physical sense a prisoner of space, is confined in a jail without walls.

216